*Improving Production
with Lean Thinking*

Improving Production with Lean Thinking

Javier Santos
Richard Wysk
José Manuel Torres

WILEY

John Wiley & Sons, Inc.

Copyright © 2006 by John Wiley & Sons, Inc. All rights reserved

Published by John Wiley & Sons, Inc., Hoboken, New Jersey
Published simultaneously in Canada

For general information about our other products and services, please contact our Customer Care Department within the United States at (800) 762-2974, outside the United States at (317) 572-3993 or fax (317) 572-4002.

Wiley also publishes its books in a variety of electronic formats. Some content that appears in print may not be available in electronic books. For more information about Wiley products, visit our web site at www.wiley.com.

Library of Congress Cataloging-in-Publication Data:

Santos, Javier.
    Improving production with lean thinking/Javier Santos, Richard Wysk, José Manuel Torres.
        p. cm.
    Includes bibliographical references and index.
    ISBN-13: 978-0471-75486-2 (cloth)
    ISBN-10: 0-471-75486-2 (cloth)
    1. Production engineering. 2. Manufacturing processes. I. Wysk, Richard A., 1948–   . II. Torres, José Manuel. III. Title.
TS176.S322 2006
658.5—dc22

2005019103

Printed in the United States of America

10 9 8 7 6 5 4 3 2 1

Contents

Preface

The paradigm of manufacturing is undergoing a major evolution throughout the world. The use of computers and the Internet has changed the way that we engineer and manufacture products. According to recent trends in manufacturing, products are subjected to a shorter product life, frequent design changes, small lot sizes, and small in-process inventory restrictions.

Computer-aided design (CAD) and computer-aided manufacturing (CAM) have become the standard for designing and manufacturing sophisticated products. Today we use CAD systems routinely to design products, and we produce them on flexible or programmable manufacturing systems (CAM). Managing manufacturing systems effectively has become as critical as using the proper engineering technology to process engineered components. Reducing waste for all aspects of engineering and production has become critical for businesses' survivability.

Improving Production with Lean Thinking is a departure from traditional production books. This book is intended for use in a course that traditionally has been titled, "Production Control," "Operations Management," "Manufacturing Systems," or "Production Management," and it is intended to provide a comprehensive view of issues related to this area, with a specific focus on lean engineering principles. This book is full of practical production examples of how lean thinking can be applied effectively to production systems.

Ever since Henry Ford pioneered manufacturing transfer flow systems and Fredrick Taylor wrote of scientific management, the world began to change by bringing high-tech consumer products into the lives of the common person. Our ability to manufacture quality products economically has inflated the standard of living throughout the world. Back in the beginning of the industrial revolution, Henry Ford doubled his worker's wages while cutting the cost of his automobile in half. This changed society forever by increasing wealth and making products more affordable. Today we have seen the same reductions in the cost of electronics hardware come about from applying good engineering and management science practice. In our global society, it is as important as ever that we use the most efficient production methods possible.

For almost a century, the United States was the world leader in automobile production. Today, however, the Toyota production system is viewed as the model for production efficiency. Interestingly, the developer of this philosophy, Taiichi Ohno, acknowledges that the stimulus for his system was his close reading of Ford's ideas. Because of this rediscovery, a new vocabulary based primarily on Japanese words to describe some of Ford's principles has found its way into all the world's manufacturing systems. Words such as *kanban, kaizen,* and *jidoka* are used routinely to describe approaches to reduce waste and make production more efficient. Mr. Ohno, Mr. Shingo, and other Japanese engineers developed a systematic approach to implement some of the good production practices that go back to the beginning of the 1900s. However, it has become far more important to systematize lean thinking because the complexity of products has increased and product life continues to get smaller and smaller.

Engineered products touch our lives everyday. Our ability to produce quality products economically affects our very standard of living. A constant focus of this book is on a systematic approach to improving production activities using lean manufacturing techniques. We feel strongly that successful managers and engineers of the future will need to understand and apply these techniques in their daily work activities. It is this area that we highlight in this book.

Unlike other production control books, this book attempts to provide a strong practical focus, along with the science and analytical background for manufacturing, improving, control, and design. This book is an excellent professional reference and also is an excellent text for instruction in both engineering and business schools.

This book comes with a companion *Instructor's Manual* that includes presentations as well as tests and examples.

Creating this book has proved that production challenges today are similar worldwide. Javier Santos and José Manuel Torres work at the University of Navarra (Spain), and Richard Wysk is professor at The Pennsylvania State University (USA). Therefore, this book includes European and American approaches to lean manufacturing issues.

This book marks the end of countless hours spent by the authors trying to refine a traditional topic into one that "hooks" to other engineering science activities. Several of our colleagues and outside reviewers read the manuscript and provided invaluable suggestions and contributions. Among them are Dr. Sanjay Joshi at The Pennsylvania State University, Dr. Matthew Frank at Iowa State University, and Bertan Altuntas. Special thanks are also due to Pablo Callejo for his artwork throughout this book. Finally, we would like to thank our families for tolerating us during the difficult parts of our writing.

Javier Santos
Richard A. Wysk
José Manuel Torres

1

Continuous Improvement Tools

Asian culture has had a significant impact on the rest of the world. Other cultures have learned and adopted many words frequently used in our daily languages related to martial arts, religion, or food.

Within the business environment, Japan has contributed greatly to the language of business with numerous concepts that represent continuous improvement tools (*kaizen tools*) and with production philosophies such as *just-in-time*. Just-in-time (JIT) philosophy is also known as *lean manufacturing*. In this first chapter, both of these production philosophies will be discussed.

Another important philosophy that will be studied in this book is the concept developed by a Japanese consultant named Kobayashi. This concept is based on a methodology of 20 keys leading business on a course of continuous improvement (*kaizen*). These 20 keys also will be presented in this chapter.

Finally, in this introductory chapter the production core elements will be presented in order to focus on improvement actions. In addition, a resource rate to measure improvement results is also explained.

CONTINUOUS IMPROVEMENT

Continuous improvement is a management philosophy based on employees' suggestions. It was developed in the United States at the end of the nineteenth century. Nevertheless, some of the most important improvements took place when this idea or philosophy arrived in Japan. Japan was already using tools such as quality circles, so when Japanese managers combined these two ideas, *kaizen* was born.

Before embarking onto *kaizen,* it is important to remark first about a contribution from Henry Ford. In 1926, Henry Ford wrote:

> To standardize a method is to choose out of the many methods the best one, and use it. Standardization means nothing unless it means standardizing upward.
>
> Today's standardization, instead of being a barricade against improvement, is the necessary foundation on which tomorrow's improvement will be based.
>
> If you think of "standardization" as the best that you know today, but which is to be improved tomorrow—you get somewhere. But if you think of standards as confining, then progress stops.

Creating a usable and meaningful standard is key to the success of any enterprise. It is not the solution but is the target on which change can be focused. Using this standard, businesses usually use two different kinds of improvements: those that suppose a revolution in the way of working and those that suppose smaller benefits with less investment that are also very important.

In production systems, evolutionary as well as revolutionary change is supported through product and process innovations, as is shown in Fig. 1.1.

The evolution consists of continuous improvements being made in both the product and the process. A rapid and radical change process is sometimes used as a precursor to *kaizen* activities. This radical change is referred to as *kaikaku* in Japanese. These revolutions are carried out by the use of methodologies such as process reengineering

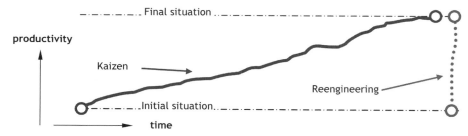

Figure 1.1. The concept of continuous improvement versus reengineering.

and a major product redesign. These kinds of innovations require large investments and are based, in many cases, on process automation. In the United States, these radical activities frequently are called *kaizen blitzes.*

If the process is being improved constantly, as shown in Fig. 1.2 (continuous line), the innovation effort required to make a major change can be reduced, and this is what *kaizen* does (dotted line on the left). While some companies focus on meeting standards, small improvements still can be made in order to reduce these expensive innovation processes. Hence innovation processes and *kaizen* are extremely important. Otherwise, the process of reengineering to reach the final situation can become very expensive (dotted line on the right).

This book presents several continuous improvement tools, most based on *kaizen,* which means improvements from employees' suggestions. As a result, all employees are expected to participate.

IMPROVEMENT PHILOSOPHIES AND METHODOLOGIES

In order to improve (quality, cost, and time) production activities, it is necessary to know the source of a factory's problem(s). However, in order to find the factory's problem, it is important to define and understand the source and core of the problem. Here it is critical to note that variability in both quality and productivity are considered major problems.

Any deviation from the standard value of a variable (quality and production rate) presents a problem. It is necessary to know what the variable objective is (desired standard) and what the starting situation (present situation) is in order to propose a realistic objective. There are three main factors that production managers fear most: (1) poor quality,

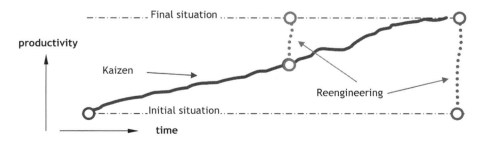

Figure 1.2. Continuous change can offset the expense and time required for radical changes.

(2) an increase in production cost, and (3) an increase in lead time. These three factors are signs of poor production management. Production improvements should be based on improvements to processes and operations. In a production area, problems can appear in any of the basic elements that constitute the area, as shown in Fig. 1.3.

Some problems, just to list a few examples, are defects, obsolete work methods, energy waste, poorly coached workers, and low rates of performance in machines and materials. By analyzing the production management history, several improvement approaches can be identified. Two of the best known improvement approaches have been chosen as references for this book: *just-in-time methodologies* (also known as *lean manufacturing*) and the *20 keys to workplace improvement* developed by Kobayashi.

Both approaches are Japanese, and their success has been proven over the last several years. The keys to the Japanese success are

- Simple improvement methodologies
- Worker involvement and respect
- Teamwork

Both these approaches are explained briefly below.

JUST-IN-TIME (JIT)

In accordance with this philosophical principle, nothing is manufactured until it is demanded, fulfilling customer requirements: "I need it today, not yesterday, not tomorrow." Only in an extreme situation, such as a product withdrawal, would it be necessary for another product to be manufactured.

The plant flexibility required to respond to this kind of demand is total and is never fully obtained. Today, it is critical that inventory is minimized. This is especially critical because product obsolescence can make in-process and finished goods inventories worthless.

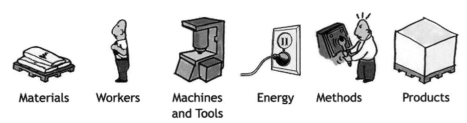

Materials   Workers   Machines   Energy   Methods   Products
                        and Tools

Figure 1.3. Resources that must be managed effectively.

In 1949, Toyota was on the brink of bankruptcy, whereas in the United States (thanks to Henry Ford's invention), Ford's car production was at least eight times more efficient than Toyota's. The president of Toyota, Kiichiro Toyoda, presented a challenge to the members of his executive team: "To achieve the same rate of production as the United States in three years."

Taiichi Ohno, vice president of Toyota, accepted his challenge and, inspired by the way that an American supermarket works, "invented" the JIT method (with the aid of other important Japanese industrial revolutionary figures such as Shigeo Shingo and Hiroyuki Hirano).

Ohno and Shingo wrote their goal: "Deliver the right material, in the exact quantity, with perfect quality, in the right place just before it is needed." To achieve this goal, they developed different methodologies that improved the production of the business. The main methodologies are illustrated in Fig. 1.4.

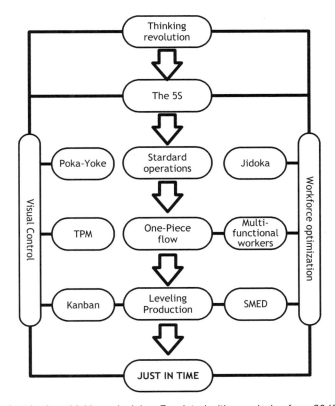

Figure 1.4. Just-in-time thinking principles. Reprinted with permission from *20 Keys to Workplace Improvement*. English translation copyright © 1995 by Productivity Press, a division of Kraus Productivity Ltd., Translated by Bruce Talbot. Appendix A translated by Miho Matsubara. Appendix C translated by Warren Smith. www.productivitypress.com.

It is important to point out that, in the figure, JIT appears as a result of several methodologies being applied, not as the beginning of a different production philosophy.

All these methodologies (besides the thinking revolution, which cannot be considered a methodology) will be studied in this book. The systematic application of all the methodologies that JIT gathered created a new management philosophy. The real value that JIT brings into the business is the knowledge acquired during its implementation. However, all these principles are not always applicable, and in several firms, some methodologies are unnecessary or even impossible to implement.

The philosophy developed at Toyota was not accepted until the end of the 1960s. Japan in 1973 benefited from the petroleum crisis and started to export fuel-efficient cars to the United States. The automobile industry in the United States decreased the cost of production and vehicle quality, but it was already too late to recover much of the automobile market. Since the 1970s, Japan has been the pioneer of work improvement methodologies.

Thinking Revolution

In the years when the JIT philosophy was being developed, the Western world employed the following formula to obtain the price of a product:

$$Price = cost + profit$$

In this formula, if the cost increases, the best way to maintain the same profit is by raising the price while maintaining the same added value in the product.

Japan, mainly at Toyota, employed the following expression:

$$Profit = price - cost$$

In this case, if the market fixes the price of a car, the only way to obtain profit is by reducing the cost. Today, this formula is used worldwide, but many years ago it was a revolutionary way of managing a company.

In order to make sure that Toyota would work like a supermarket filled with perishable goods that cannot be held too long, a new philosophy was adopted. When a product is withdrawn, the system must be able to replace it in a short period of time so that the system will not "starve." To accomplish this, it was necessary to identify and eliminate in a systematic way all business and production wastes.

Seven Types of Waste. At Toyota, management follows the principle that the real cost is "as big as a seed of a plum tree." One of the main problems in production management is to identify cost's true value.

In some cases, manufacturers let the seed (cost) grow as big as a tree. Unfortunately, the greater the cost, the greater is the effort required to decrease it. This can be compared with the fact that managers try to decrease cost by cutting some leaves out of the growing tree to improve the factory. This means that cutting the leaves from a tree improves the tasks that add value to the product.

In reality, it is more efficient to eliminate tasks that do not add value to the product. Reducing the tree to a smaller size is equivalent to planting a smaller seed and not letting it grow. In other words, finding the real production cost can be difficult but is necessary.

The goal of Toyota's executives was to find this plum tree seed and work hard to reduce cost until it reached the size of the seed just mentioned, not allowing the cost to grow into a leafy tree. In order to achieve this goal, they needed to eliminate all tasks that did not add any value to the process and thus leading to cost increases.

Hiroyuki Hirano defined *waste* as "everything that is not absolutely essential." This definition supposes that few operations are safe from elimination, and this is essentially what has happened. He also defined *work* as "any task that adds value to the product." Toyota's factories outside Japan required between 5 to 10 times more operations to produce the same car as its Japanese factories. The elimination of waste and the decrease in production inefficiencies rapidly convinced managers that this philosophy was going to be successful.

In conclusion, it was possible to realize the goal by changing work methods instead of attempting to do the operations at a faster speed.

Shigeo Shingo identified seven main wastes common to factories:

- *Overproduction.* Producing unnecessary products when they are not needed and in a greater quantities than required.
- *Inventory.* Material stored as raw material, work-in-process, and final products.
- *Transportation.* Material handling between internal sections.
- *Defects.* Irregular products that interfere with productivity, stopping the flow of high-quality products.
- *Processes.* Tasks accepted as necessary.

- *Operations.* Not all operations add value to the product.
- *Inactivities.* Machines with idle time or operators with idle time.

Of all these types of waste, inventory waste is considered to have the greatest impact. Inventory is a sign of an ill factory because it hides the problems instead of resolving them, as shown in Fig. 1.5.

For example, in a factory, in order to cope with the problem of poor process quality, the size of production lots typically is increased. As a consequence, products that probably will never be used get stored. If the problem that produces the low quality is solved (equivalent to breaking the rocks in the figure), inventory could be reduced without affecting service.

Sometimes, because of resistance to change, the inventory level does not decrease after the improvement. In such cases it will be necessary to force a decrease in inventory (this is equivalent to opening the dam's door in the figure).

In addition, holding cost (the cost to carry a product in inventory) frequently is underestimated. The maintenance and repair costs of the inventory equipment or material handling elements are not usually considered.

Lean Manufacturing

Basically, *lean manufacturing* is the systematic elimination of waste. As the name implies, *lean* is focused on cutting "fat" from production

Figure 1.5. Inventory can hide production inefficiencies and slow improvements.

activities. *Lean* also has been applied successfully to administrative and engineering activities. Although *lean manufacturing* is a relatively new term, many of the tools used in lean manufacturing can be traced back to Fredrick Taylor, Henry Ford, and the Gilbreths at the turn of the twentieth century. The Japanesse systemitized the development and evolution of improvement tools.

Lean manufacturing is one way to define Toyota's production system. Another definition that describes lean manufacturing is *waste-free production*. *Muda* is the term chosen to refer to lean manufacturing. In Japanese, *muda* means waste. Lean manufacturing is supported by three philosophies, JIT, *kaizen* (continuous improvements), and *jidoka*.

Jidoka is a Japanese word that translates as "autonomation," a form of automation in which machinery automatically inspects each item after producing it, ceasing production and notifying humans if a defect is detected. *Jidoka* will be explained in Chap. 9.

Toyota expands the meaning of *jidoka* to include the responsibility of all workers to function similarly, i.e., to check every item produced and to make no more if a defect is detected until the cause of the defect has been identified and corrected.

According to the lean philosophy, the traditional approximations to improve the lead time are based on reducing waste in the activities that add value (AV) to the products, as is shown in Fig. 1.6.

Lean manufacturing, however, reduces the lead time by eliminating operations that do not add value to the product (*muda*). According to lean manufacturing, lead time should not be 10 times greater than the added-value time (time that adds value to the product), as is shown in the Fig. 1.6 on the right.

When the *lean team* is established, and if the team operates effectively, the most important wastes are detected and eliminated.

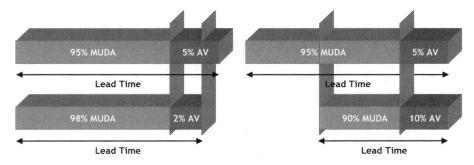

Figure 1.6. Saving time means eliminating waste.

20 Keys to Workplace Improvement

Iwao Kobayashi, in 1988 published a book explaining 20 keys to workplace improvement. They all must be considered in order to achieve continuous improvement.

These 20 keys are arranged in a circle (Fig. 1.7) that shows the relations between the keys and their influence on the three main factors explained previously: quality, cost, and lead time. The arrangement in the circle is not categorical, and some keys offer benefits in more than one factor.

There are four keys outside the circle. Three of them (keys 1, 2, and 3) must be implemented before the rest, and key 20 is the result of implementing the other 19 keys.

Figure 1.7. The 20 keys to workplace improvement. Reprinted with permission from *JIT Factory Revolution: A pictorial guide to factory design of the future*. English translation copyright © 1988 by Productivity Press a division of Kraus Productivity. Ltd. www.productivitypress.com.

Kobayashi divided each key into five levels and set some criteria to rise from one level to the next. The first step in the methodology consists of specifying the actual company's current level and then the required level. After figuring out the current level of the company, Kobayashi offers the steps the company must use to reach the final level gradually rather than attempting to reach the top directly (Fig. 1.8).

On the other hand, to show the evolution of the factory, Kobayashi presents a radar graphic (Fig. 1.9) in which the scoring of each key is represented.

Kobayashi recommends improving all the keys equally. Because of this recommendation, in the radar graphic, the factory's scoring will grow concentrically.

MEASURING AND PRIORITIZING THE IMPROVEMENTS

Today, no one questions the utility of these methodologies: They have been implemented successfully in several companies. Nevertheless, there are problems in prioritizing the importance of an implementation, as well as problems in the way that increased improvements are measured. In this book, a classification of improvement methodologies is presented based on a known production rate: overall equipment efficiency.

Overall Equipment Efficiency. To improve the productivity of production equipment, it is necessary to know the actual equipment state by analyzing its component activities. Nakajima summarized the main time losses for equipment based on the value of three activities.

Figure 1.8. Assessing the current position (level) and the target position is critical to success.

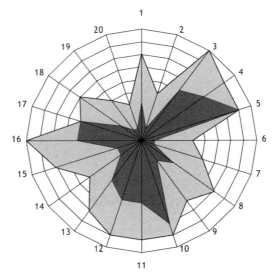

Figure 1.9. Radar graphic for each factor.

Considering the available work time, referring to the calendar time, there is a fixed time for planned stops: preventive maintenance, operators break, etc. (Fig. 1.10). The rest of the time is considered *load time* (or *machine load time*).

Load time can be reduced based on the six main causes for a reduction in valid operating time. The causes for losses that affect a machine's productivity include:

- *Breakdowns,* referring to, the time that the machine is stopped by repairs.
- *Setup and changeovers,* which correspond to the change time between models or between products of the same model.
- *Idling and minor stoppage,* referring to loss time caused by the processes' randomness or by the worker-machine cycle complexity.

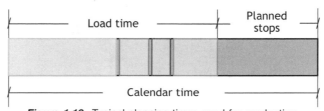

Figure 1.10. Typical planning times used for production.

- *Reduced speed,* caused by the wear of components.
- *Defects and reworks,* referring to low-quality products.
- *Start-up losses,* because the machine produces defects until it reaches the operation steady state.

Figure 1.11 illustrates how these six main losses are grouped. Each cause is analyzed in order to reduce the load time until the real useful time of the equipment (the real operating time of the equipment) is reached.

In addition, the preceding grouped losses define three basic indicators: availability, performance, and quality. Measurements for these losses (expressions) are presented in Fig. 1.12.

Finally, in the same figure, the expression for the *overall equipment efficiency* (OEE) also is shown. This is the rate that includes all the losses that a piece of production equipment can have and also allows the prioritization of improvement actions.

The objectives predicted for each indicator by Nakajima are more than 90 percent in the availability, more than 95 percent in the rate of performance, and more than 99 percent in the rate of quality. However, the main advantage of implementing these rates, established by Nakajima, is that they can show how the improvements carried out affect the equipment efficiency directly.

Figure 1.13 shows different impacts on the equipment efficiency rate caused by a maintenance improvement project. The figure also shows the starting situation in order to allow comparison of the different re-

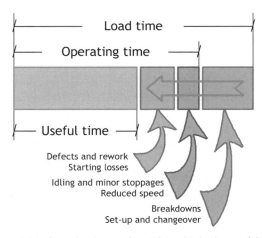

Figure 1.11. Grouping losses (waste) to obtain the useful time.

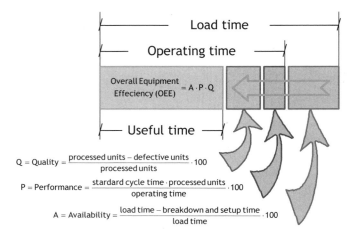

$$\text{Overall Equipment Efficiency (OEE)} = A \cdot P \cdot Q$$

$$Q = \text{Quality} = \frac{\text{processed units} - \text{defective units}}{\text{processed units}} \cdot 100$$

$$P = \text{Performance} = \frac{\text{stardard cycle time} \cdot \text{processed units}}{\text{operating time}} \cdot 100$$

$$A = \text{Availability} = \frac{\text{load time} - \text{breakdown and setup time}}{\text{load time}} \cdot 100$$

Figure 1.12. Measuring losses in a system.

sults achieved. Maintenance improvement can produce three types of situations:

- Transitory improvement
- Permanent improvement
- Permanent improvement (shown in the availability rate increase) but worsening of the OEE rate

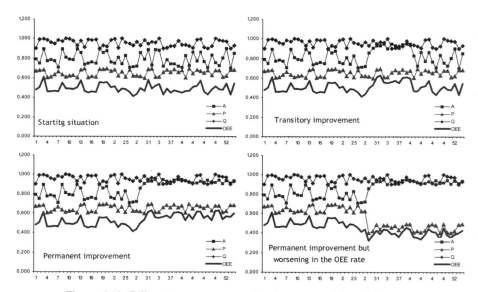

Figure 1.13. Different consequences of an improvement in maintenance.

Each indicator usually is represented in an independent graph to facilitate its reading and detailed analysis. Besides, they usually present generally similar values, and the graphs could mix.

BOOK STRUCTURE

The rest of this book is divided in seven chapters dedicated to the improvement of different aspects of the production area. In these chapters, well-known tools and improvement methodologies are defined and illustrated.

The structure of all chapters is similar, although some material could be omitted:

- *Introduction,* in which the particular subject and its relation to other sections is presented.
- Presentation of the needed *theoretical bases.*
- Explanation of the *methodology* that enables the proposed improvement
- *Support tools* study for all methodology steps
- Itemization of the expected *effects and benefits* after the methodology is applied
- Presentation of *recommended readings* to increase and expand knowledge
- *Flow of materials.* Methodologies oriented to reduce the material movements (Chaps. 2 and 3).

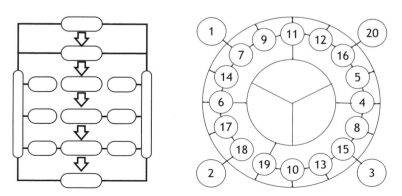

Figure 1.14. Icons that will represent each methodology in the JIT and 20 keys schemas.

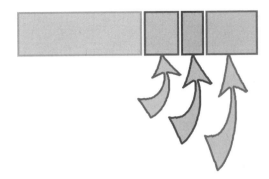

Figure 1.15. Impact of the methodology on the OEE rate.

- *Efficiency of the equipment.* Methodologies focused on increasing the OEE value explained previously (Chaps. 4, 5, 6, and 7).
- *Work environment.* This is based on the 5S methodology. This methodology allows the work environment to be prepared and made ready for the other methodologies to be established. Therefore, this chapter is placed at the end on this book (Chap. 8).

In addition, there is a final chapter (Chap. 9) where other improvement tools are explained briefly.

Finally, the methodology studied will be situated in the JIT and 20 keys schemas (Fig. 1.14).

In the second part, where the efficiency of the equipment is analyzed, the improvement impact on the relevant OEE rate will be represented by the chart in Fig. 1.15, including only the arrows that will be affected by the improvement.

RECOMMENDED READINGS

Henry Ford, *Today and Tomorrow.* Productivity Press, New York, N.Y. 1988.

Hiroyuki Hirano, *JIT Implementation Manual,* Vols. 1 and 2. Productivity Press, Portland, OR 1990.

Hiroyuki Hirano, *Factory Revolution.* Productivity Press, Portland, OR 1989.

Iwao Kobayashi, *20 Keys to Workplace Improvement,* revised edition. Productivity Press, Cambridge, MA 1998.

Yasuhiro Monden, *Toyota Production Systems,* 3d ed. New York: Springer, 1998.

Seiichi Nakajima, *Introduction to TPM: Total Productive Maintenance.* Productivity Press, Portland, OR 1995.

William J. Stevenson, *Production and Operations Management,* 5th ed., New York: Irwin McGraw-Hill, 1996.

2

Material Flow and Facilities Layout

Several productivity metrics, such as throughput and lead time, are directly affected by the where and how the processing and storage resources are located in a factory. In this chapter, different types of industrial processes and plant layouts are discussed.

Plant layout (changes in resource or even factory location) is an activity that all companies are forced to deal with sooner or later. These situations occur because of technology innovations, increases in demand, and certain other productivity reasons. Therefore, it is important to be familiar with the methodologies used to carry out these kinds of studies.

Cellular layouts—where labor and machines are grouped in cells—will be presented in this chapter because they are becoming increasingly important and also because they require specific methodologies. Cellular layouts will be explained in greater depth in Chap. 3.

LAYOUT IMPROVEMENTS

Factory layout improvements typically occur more than one time during a factory's life. The study of plant layouts seeks the optimal location for all the production resources. At the same time, the study tries to ensure that the economic impact of the project on the enterprise will be as positive as possible. Lastly, the new plant layout must be as safe as possible and satisfactory for the employees.

18

With all these restrictions, it seems obvious that the optimal solution could be unreachable. In reality, the ideal plant layout can be reached if and only if a commitment between all the aspects previously mentioned is achieved.

Signs and Reasons for a Need to Change the Layout

The signs and reasons to propose a change in a plant layout are varied, but some of these reasons appear more frequently and therefore are described below.

Location Change. There are multiple reasons to suggest a change of location of a factory. Some companies were founded many years ago in locations that have become small or antiquated. Some companies are located in an urbanized area where a factory extension is impossible.

The starting point for a new plant layout is also different if the company chooses a new location and carries out new building construction or if the new location is already built. Today, new building constructions allow for an ideal layout because, with a few exceptions, building functionality is the principal focus in the building's design. As a result, the factory surface is better utilized.

Purchase of New Equipment. New needs and technology improvements form the basis of machinery purchase. Finding the best location for the purchased equipment can become a critical issue in making a "system" perform as intended.

Newly purchased incremental equipment generally is placed in the first free space available. In some cases, when there is no free factory space, it is necessary to move machines to create space. In other cases, a global layout project is outlined, and the new machine is located in a place that promotes system efficiency. These equipment movements can be crucial in production effectiveness as well as in the consideration of future equipment purchases.

Problems with the Materials Flow. This problem generally derives from the problem presented previously: placing new equipment in the first available corner of a plant. As a result, the initial setup costs decrease, but other problems arise later. Depending on the operation that the new machine carries out and its relative situation with respect to the preceding and next machine (in terms of process flow), the materials flow can be adversely affected on a larger or smaller scale.

It is important to remember that equipment setup is done only once, whereas materials flow is a continuous process. The analysis of this materials flow justifies the investment in time and money needed in the new equipment incorporation study and in most cases can be demonstrated economically. It is recommended to analyze an optimal location for the new equipment that will improve materials flow even before buying the equipment.

High Work-in-Process (WIP) Most company situations vary over time, so what may have been a good policy or layout in a given period of time may not produce good results forever. A measure or good indicator of change in a company is the amount of partially completed products (work-in-process). It is important not to confuse a temporary situation caused by a momentary increase in demand with a permanent and untenable situation.

It is only in the case where product mix and batch size changes occur that a company proceeds to a detailed layout analysis. Slow changes in product variability can hide the negative effects of the excessive work-in-process (WIP) caused by engineering and demand changes in the manufactured products.

THEORETICAL BASIS

One-Piece Flow

Before we start explaining the layout analysis tools, it is important to clarify the definition of *production* and *transfer batch* even though both sizes normally are the same:

- *Production batch* refers to the number of products included in a customer order.
- *Transfer batch* is defined as the number of units that flow from one machine to the next machine. This is also referred to as a *unit load.*

The work-in-process decreases according to the reduction in transfer batch size, as shown in Fig. 2.1.

In the first case, the transfer batch is equal to the production batch, whereas in the second case, the number of pieces that flow is one-third the production batch. The advantages of reducing the transfer batch are

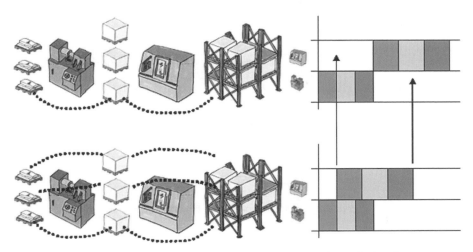

Figure 2.1. Difference between production batch and transfer batch.

- *Production feedback is faster.* As a result, information about the quality of a product is received sooner, and the reaction to prevent rejection of the whole batch is quicker.
- *The lead time decreases.* This was shown in Fig. 2.1.

However, reduction in the transfer batch increases the materials handling between sections in the same proportion as the batch gets decreased.

The ideal transfer lot size, and therefore, the ideal flow between working areas, is what is called a *continuous one-piece flow*. This definition outlines one problem: If one of the production areas stops, the whole production plant will stop owing to the absence of material. This problem forces the number of pieces or work-in-process to be established beforehand. Normally, a container size is considered one-piece flow. For example, a "1000-screws flow" can be used as one-piece flow.

One-piece flow eliminates most of the signs and reasons for a need to change the layout outlined in the preceding section, and it is one of the just-in-time (JIT) tools (Fig. 2.2).

To come closer to an ideal one-piece flow, the material flow between both equipment and workstations has to be minimized or eliminated (directly linking process activities). If this is not possible, then the machines should be located as close together as possible. The bottom line is that in order to improve the material flow, it is typically necessary to analyze and change the company's layout.

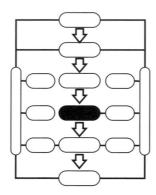

Figure 2.2. Location of one-piece flow in JIT schema.

Main Types of Industrial Companies

In the market there exists a multitude of different products: food, cars, computers, bricks, cement, ships, etc. Each product has a specific manufacturing process. The study of all these processes would require a heavy investment in time if they are not grouped by similarities.

The manufacturing processes of a car and a washing machine are similar, and they could be analyzed in the same manner. They can be considered "similar products." In the same way, the manufacturing processes of yogurt and soap also can be studied together. However, no one would eat a conditioner bottle for dessert or wash their clothes with a yogurt shake.

Product grouping allows for the application of the same principles and tools to seemingly different companies, basing the groupings on the types of production facilities the companies use. As a result, all companies considered *process industry* have similar plant layout functionality (yogurt and soap production belong to this kind of industry).

Industrial companies, as well as the economy, can be grouped into four sectors: primary, secondary, tertiary and service sector. The primary sector and the service sector will not be considered in this book. Study will be centered on the secondary sector (process industry and consumption industry), as well as on the tertiary sector (production and assembly factories).

Process Industry. Several industries are process-focused industries, where manufacture of the product dictates the equipment and product flow. Included in this type of manufacturing are paper, wood, cement, painting, and fabrics manufacturers.

Some factories that produce consumption goods (e.g., beverages), although not properly considered part of the process industry, also will be grouped into this type because the manufacturing process is configured very similarly, although, logically, the hygiene measures will be much higher. Consumption goods are all food products such as yogurts, ice creams, and drinks, as well as pharmaceutical and cleaning products.

The typical process industry, as shown in Fig. 2.3, has four main steps:

- *Raw materials preparation.* The main raw materials usually are received in bulk and are stored in large warehouses or silos. The product mixing (or formula preparation) is carried out in hoppers, blenders, or smaller deposits. The exact material quantities are obtained from an ingredients list and then completed using scales and measures for the specific formulas for each article.
- *Treatment.* Depending on the product type or its specifications, treatment (or processing) is carried out through different operations, such as filtering, drying, sieving, etc.
- *Finishing.* Metals, for example, can lose some properties during the treatment process; therefore, it is necessary to restore them. In other cases, superficial treatments such polishing or painting are applied to obtain the final aspect of the product.
- *Bottling or packaging.* Finished products will go through the bottling or packing lines depending on their needs. This is the case for either chips or wine. Concrete, for example, is dumped in trucks that will transport it to the corresponding building or con-

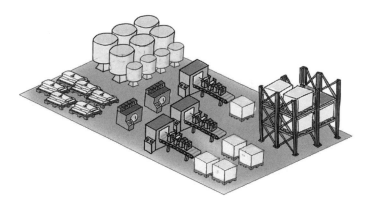

Figure 2.3. Example of a process industry.

struction area. A product such as soap is packed in packages or bags of different sizes.

It is important to clarify that some companies will have their own nuances, although they will be studied in the same grouping. For example, in a yogurt manufacturing company, after the various yogurts are packed, they must be fermented, whereas in a candy factory, the final products simply are stored.

Assembly Companies. Most of the products that are used daily (e.g., cars, televisions, microwaves, etc.) are manufactured in companies that exclusively assemble final products, buying most of the assembly components (Fig. 2.4).

These companies have assembly lines where the products or families of products are produced. Their components are purchased from external companies because these companies are responsible solely for the assembly. In some cases, e.g., in automobile or electrical appliance companies, some processing operations are also carried out within the factory (e.g., sheet cutting, welding, plastic injection molding, or painting). These are isolated operations performed by these companies because they are not profitable to subcontract.

Manufacturing Companies. Companies that manufacture component parts do not belong to any of the preceding groups. These companies are called *manufacturing companies* and are known for

Figure 2.4. Motorbike assembly line.

processes such as forges, plastic injection machines, presses, computer numerical control (CNC) machines, etc.

Factory layout for manufacturing companies depends on the product type and volume to be manufactured. Later on in this chapter this type 8 layout will be analyzed in more detail. Figure 2.5 presents a tradition process layout example of a manufacturing company that specializes in the production of component parts.

Layout Types

There are numerous classifications of industries based on layout. We will use four basic layout groupings or classifications. The grouping is primarily the result of the materials flow in the production plant.

- Fixed-position layout
- Process layout
- Product layout
- Cellular or combination layout

Fixed-Position Layout. In this type of layout, the product does not move throughout the production process; the needed resources move (Fig. 2.6). This layout is used in the manufacture of products that are difficult to move (e.g., ships, buildings, trains, etc.) or in products with short or immediate needs (e.g., milling center, presses, etc.).

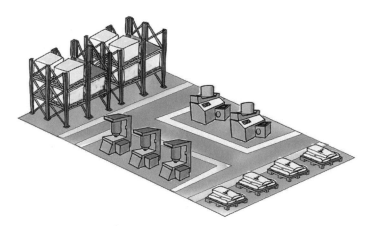

Figure 2.5. Example of a manufacturing company.

Figure 2.6. Fixed-position layout.

Historically this layout also was used for car production, although today the way automobiles are manufactured has changed drastically.

Process Layout. In this type of layout, machines are grouped into departments or stations according to the operations they perform. For example, presses are grouped in the pressing department (Fig. 2.7), and lathes form the lathe department.

This layout is used commonly in companies that manufacture by orders for specialty parts or components (one or a few of a kind). For example, a small job shop that makes unique dies or fixtures would use a process layout. Process layout typically is employed for a large variety of products that are made in very small batches (ones or twos).

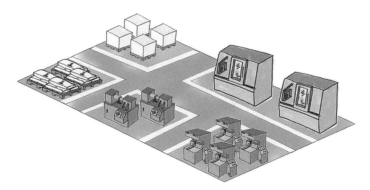

Figure 2.7. Example of a process layout.

The advantages of a process layout are

- The system has the flexibility to produce almost any part that fits within the volumetric boundaries of the machines.
- An in-depth understanding of a specific process can be obtained.
- Some tooling and fixtures can be shared.

The disadvantages of process layout are

- The spaghetti flow is difficult to manage and control.
- There is usually a lot of inventory in front of each machine.
- Setup is usually expensive.
- Materials handling times are large.
- It is difficult to automate these types of systems.

Product Layout. In this type of layout, the machines are grouped according to the product manufacturing sequence (Fig. 2.8). Depending on the main activity of the production line, these layouts are called *manufacturing* or *assembly lines.* High volume component parts normally are produced using a product layout.

Assembly companies normally use this layout, especially in the automotive sector (Fig. 2.9). The layout change carried out by Henry Ford drastically reduced car production lead time. Today, some companies are able to manufacture an automobile every 40 seconds.

Product layout systems are used effectively for the economic production of high-volume goods. The advantages of these systems are

- Large batches can be produced inexpensively.
- Materials handling is minimal.
- In-process materials are minimized.

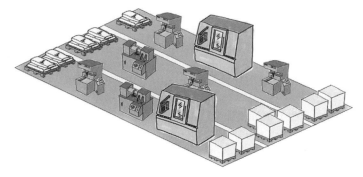

Figure 2.8. Example of a product layout.

Figure 2.9. Assembly companies.

- It is easy to control these systems.
- Automation is more achievable and justifiable.

The disadvantages of these systems are

- They are inflexible, in that only one or very few products can be produced on them.
- Setup time for these systems is very large.
- Duplicate tooling is required to replace worn tooling so that maintenance can be minimized.

Cellular or Combination Layouts.   Some companies cannot be classified exclusively as having one of the preceding layout types. The following paragraphs present some examples.

Large products manufacturing industries such as airplanes and presses have opted for modularization as the best way to simplify product assembly tasks. As a result, to assemble a large CNC machine, different modules are produced in different lines and assembled as subsets (Fig. 2.10).

Many companies, such as special screw manufacturers, owing to the demand, have been forced to change their layouts, dismantling sections and creating manufacturing cells (which will be discussed in Chap. 3). *Cellular layout* is a relatively new layout approach and will be treated separately from traditional layouts.

In some other companies interested in creating manufacturing cells, it is not possible to purchase or have available all the necessary ma-

Figure 2.10. Modularization in machine production.

chines for the high cost that this action implies. In this case, the company modifies the layout in order to allow all cells to share the critical resource. This is a *combination layout* between product and process layouts.

Other companies have a common first phase, organized according to the process layout, and different assembly lines to elaborate the final product. This distribution type is seen commonly in assembling companies, such as the appliances manufactures, with plastic injection and presses sections combined with assembly lines grouped by product families.

Characteristic of the Traditional Layouts

The three traditional layouts explained (fixed-position, process, and product layouts) have specific characteristics that make them suitable for some companies and not very appropriate for other types of companies. Table 2.1 presents a summary of the main ideas explained in preceding sections.

LAYOUT DESIGN METHODOLOGY

Six basic steps are necessary to design an acceptable solution for a layout problem. Although these steps are applicable to most layout problems, they are oriented mainly for a general layout analysis. If the study objective is more restricted, it is not necessary to apply all the steps.

TABLE 2.1.   Main Characteristics of the Three Traditional Layouts

	Fixed Position Layout	Process Layout	Product Layout
Product	Difficult to move or with small and specific demand	Products diversified with variable production volume	Standard products with high production volume
Material flow	The product does not move	Manufacturing- particular path (standard routes do not exist)	Unidirectional and the same one for all products
Machinery	General machinery and common to all products	Each machine manufactures different products	Specific machinery for each operation
Labor	The task assignment depends on the project	Specific skill in each process	Repetitive tasks, although rotation of the staff is favored

Step 1: Formulate the Problem

Although it seems trivial, it is extremely important to define what the main objective of the study is: Including a new machine? Modifying the existing building? etc.

Step 2: Analysis of the Problem

Analysis of the actual situation can be carried out in a systematic way. Richard Muther in his classical book, *Practical Plant Layout,* presented eight factors to consider for facilities layout. These factors will be described briefly in the tools section of this chapter.

Step 3: Search for Alternatives

Analysis of Muther's eight factors enables engineers to define the problem and align the solution properly to the problem. Nevertheless, it is important to take into account three practical principles.

First the Whole and Then the Details. Both large layout changes and cellular layout design problems should be kept in mind, giving priority to the general area or total space shared and then to each one of the specific areas (Fig. 2.11). Layered planes are developed to characterize the situation. The layered planes help to illustrate the general flows between different departments.

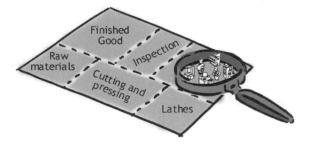

Figure 2.11. First the whole and then the details.

First the Ideal Solution and Then the Practical One. It is important to realize from the very beginning of the project that the ideal solution is difficult to reach (Fig. 2.12). However, in many cases good solutions, very near to the ideal solution, can be developed more easily.

Realizing that the ideal solution requires infinite knowledge, a more particle approach is more worthwhile so as not to waste time and effort analyzing the problem in depth.

Brainstorming. The first step in the layout process is the *creation of ideas,* where several possible solutions are generated, with not a single solution being rejected. The brainstorming methodology recommends considering all the ideas the development team proposes without criticizing them. Being critical at this point can hinder the creativity process, and sometimes ideas that appear "far out" at first become realistic solutions with small changes. It would not be the first time that brilliant solutions are obtained from a seemingly crazy idea.

It is also important to remember that factories have a third poorly used dimension: the height or overhead space. Today, a significant

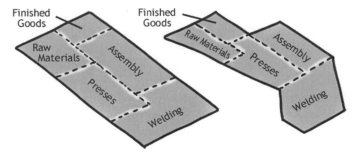

Figure 2.12. First the ideal solution and then the practical one.

number of companies use overhead space as temporary storage. As a result, the useful area of the plant is increased.

Step 4: Choose the Right Solution

The objective of this step is to choose the solution that fits best among the solutions that have been proposed in the preceding step. Each one of the solutions should be evaluated according to a specific set of criteria.

A simply way of evaluating these possible solutions consists of ranking each alternative from 0 to 10 according to established criteria. The solution that obtains the best overall ranking will be the accepted one.

It is also very important to evaluate each alternative from an economic standpoint because frequently it is the "money factor" that determines which path is taken. The advantages and disadvantages of each solution should be specified, and always keep in mind that the simplest solution (between those which have received good rankings) will be the best choice. Many company folders contain plenty of completed and detailed layout studies that propose radical changes in the current configuration and that were never put into practice.

Step 5: Specification of the Solution

The accepted solution will need to be fully developed. Many details are not considered or defined in the preceding step because it is pointless to present a complete solution if it is not the definitive one.

This step is also useful to take care of safety measures in order to avoid possible future industrial accidents. Occupational Health and Safety Administration (OHSA) regulations and the Labor Risks Prevention Law provide guidelines for some minimum working solutions, e.g., minimum worker access. The adopted solution must be consistent with all laws and regulations. An example is shown in Fig. 2.13, where the situation presented on the left side of the figure should be avoided, migrating toward solutions such as the one presented on the right side.

Finally, it is necessary to itemize all details of the plan, request the corresponding budgets, and establish a schedule for implementation of the solution. It is also important to demonstrate quantitatively that the outlined solution will provide benefits when compared with the current situation.

Figure 2.13. Labor safety is a must in the outlined layout.

Step 6: Design Cycle

The design cycle includes planning for modifications that arise because of problems that appear while adopting a solution: budget deviations and/or problems in plant installations (electric or pneumatic lines).

At the end of the design process, the plant should work more efficiently. It is always worthwhile to check to see if the adopted solution works as expected.

TOOLS FOR LAYOUT STUDY

Muther's Eight Factors

By analyzing the following eight factors, it is possible to determine the main layout restrictions and requirements that a new layout alternative has in order to choose the best layout from a set of proposed solutions.

Material Factor. The material factor does not cover study of the materials used to manufacture the product, although the name might imply that it should. The purpose of this factor is to become familiar with the different production steps needed to manufacture the article and to analyze how the material is transformed from raw material to a final product.

The product sequence of operations should be studied without considering the relative location of each process in the factory. For example, to manufacture a screwdriver, the fist step consists of transforming a steel bar into the screwdriver bar with the right tip. The handle then would be created using an injection-molding process for a plastic handle, with the two pieces then being assembled. This factor helps you to understand the company technology and to know the company's range of products.

Machinery Factor. The second factor analyzes the machine types and the existing number of machines on the factory floor. It is important to take note of the number of each type and their principal dimensions in case this becomes a critical constraint. This machinery factor is shown in Fig. 2.14.

It is also necessary to analyze the operating conditions, such as vibration, temperature, etc., so as to avoid putting incompatible machines together. For example, a heavy sheet metal press and a precision coordinate measurement machine are not very compatible.

Labor Factor. The staff of the production department should be counted, from machine operators to section heads. In addition, materials handling and maintenance operator input is important.

To facilitate study of this factor, we recommend using worker-machine diagrams (a tool that will be explained in Chapter 5). This tool allows you to discover the operations that workers carry out on the machines and the relative disposition of the elements in the work area so as to simplify worker tasks.

Movement Factor. The movement factor analyzes the materials flow between work centers. This flow does not add value to the product, and as a consequence, as much handling as possible should be eliminated. Logically, completely eliminating movement is an ideal, but most often it is feasible to eliminate certain handling components to

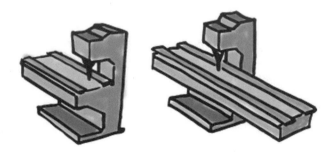

Figure 2.14. Considering machine dimensions.

obtain a better solution. There are mainly two tools to analyze movement between machines: the flow-process chart and the transfer matrix.

Flow-Process Chart. A *flow-process chart* represents, in a graphic way, the path and the actions carried out on a product. Five standard symbols are used to show all the alternatives (Fig. 2.15). Also, it is possible to combine two or more of these actions, creating two or more new symbols.

The circle symbol represents an operation, the arrow symbol represents a transport, the square represents an inspection, the reversed triangle represents a storage, and the letter *D* represents a wait or delay. The difference between the last two symbols is that, in the first case, it is necessary to remove the product from a warehouse or inventory location after storage.

Using these symbols as tools, movement improvements can be envisioned and advantages quantified when a modification is presented, as shown in Fig. 2.16.

Transfer Matrix. A *transfer matrix* is a matrix representation of the work flow in a production plant. The matrix shows the fraction of work that flows from one section to all the others, including the raw materials and final product warehouses (Fig. 2.17).

Preparation of a transfer matrix is not complicated. The matrix considers the total amount of product that enters a work center, and the fraction moving to other work centers is calculated. This is distributed

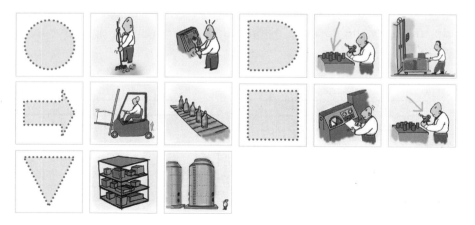

Figure 2.15. Standard symbols on a flow-process chart.

Product 1				
	Number	Time (min)	Distance (m)	Workers
Operation ⬤	3	80	-	4
Transport ⇨	5	13	102	5
Inspection ▪	-	-	-	-
Storage ▼	3	2	-	2
TOTAL	11	95	102	11
Product 2				
	Number	Time (min)	Distance (m)	Workers
Operation ⬤	4	140	10	5
Transport ⇨	7	20	167	7
Inspection ▪	-	-	-	-
Storage ▼	3	482	-	2
TOTAL	14	642	177	14

Figure 2.16. Quantification of movement by means of the flow-process chart.

fractionally among the other sections; therefore, all the rows of the matrix should sum to 1.

For example, in Fig. 2.17 it can be seen that half the work (50 percent) that enters M1 is sent to M2, 30 percent to M3, 10 percent to M4, and the last 10 percent to M5.

The matrix shows the volumes of different products that flow between work centers, but it does not indicate how heavy they are or their size. Therefore, the information in the matrix is not enough to make a fully informed decision about the convenience of locating two work centers adjacent to each other.

As a result, it is possible to use the same matrix concept but with another perspective, keeping in mind other factors such as the transferred weight or the number of routes executed. These matrixes will help to clarify the decision of the relative locations of departments in the factory plant.

TO

		M1	M2	M3	M4	M5	W2
	W1	0.7	0.2		0.1		
F	M1		0.5	0.3	0.1	0.1	
R	M2			0.7	0.2		0.1
O	M3				0.6	0.2	0.2
M	M4					0.6	0.4
	M5						1

TO

		M1	M2	M3	M4	M5	W2
	W1	0.3		0.2	0.1	0.4	
F	M1			0.3	0.1	0.4	0.2
R	M2	0.7					0.3
O	M3	0.5			0.1	0.1	0.3
M	M4		0.2	0.2			0.6
	M5	0.1	0.1	0.1	0.6		0.1

Figure 2.17. Examples of transfer matrices.

Wait Factor. This factor covers study of the three main warehouses: raw materials, work-in-process, and final product. The objective of the wait factor is to determine the required space in each of the warehouses. Muther recommends in-depth analysis of the required space for each product. It normally happens that, owing to the magnitude of the warehouse study, a specific layout project must be outlined. Because the warehouse layout is closely related to its management (planning and control), it will not be studied in this book.

Service Factor. The service factor is used to analyze two different characteristics:

- Study of environmental workspace conditions (i.e., brightness, noises, smells, minimum working space) in order to decide what the acceptable parameters are with respect the OHSA regulations and the Labor Risks Prevention Law.
- Related with the preceding characteristic, the working conditions are analyzed with emphasis on the plant service staff. Plant services are mainly quality, logistics, and maintenance. Minimum maneuver space for forklift trucks or other special equipment used in these services is a typical factor.

Building Factor. The building factor analyzes the actual useful surface of the building. This factor takes into account the plant shape, the columns, the window situation for ventilation, and areas of possible extension. In many cases the surface area covered by gantry cranes limits the number of layout alternatives because in many cases these resources must overlap, and they cannot be moved easily.

Change Factor. Regrettably, the proposed layout will not be valid forever. Neither is it the goal of this factor to leave the company ready for any future change because the future is usually unknown.

The change factor is intended to observe, from a critical point of view, the adopted solution. For example, if the new raw materials warehouse has been designed without free space, it is very probable that the layout will have to be reconsidered in the near future if demand grows. As a consequence, the work already done will be of little value.

Application of this factor is without a doubt the most difficult part of the study. It is necessary to ask for future company plans (e.g., increases in the number of references, market target changes, etc.) with the purpose of extending the usefulness of the proposed layout for as long as possible.

SUMMARY

This chapter has demonstrated how materials flow can be improved significantly by means of layout analysis. In these kinds of improvement projects, study of the current situation allows one to identify constraints that reduce the number of possible alternatives to be considered. The proposed improvements will reduce the materials flow, allowing the company to raise the one-piece flow proposed by the lean manufacturing philosophy.

RECOMMENDED READINGS

Richard Muther, *Practical Plant Layout.* New York: McGraw-Hill, 1956.

Kjell B. Zandin and Harold B. Maynard, *Maynard's Industrial Engineering Handbook,* 5th ed. New York: McGraw-Hill, 2001.

3

Material Flow and the Design of Cellular Layouts

In Chapter 2 the concept of manufacturing cells was presented as a specific case of the product/process layout. The basic analysis necessary to transform a traditional factory into a cellular layout requires a unique development and implementation methodology. Because of this difference in layout analysis and philosophy, a separate chapter on cellular design and analysis is used for this important topic.

The use of cells creates a unique set of production modules. The division of the plant into cells exclusive to the production of a product family transforms the factory into a group of self-managed subfactories or modules. This chapter presents some design and analysis tools focused on getting a company ready to progress to cellular manufacturing.

When a production line is being designed, it is important to distribute the needed manufacturing tasks within the workstations as best as possible. One always should avoid any unneeded workstations so that task distribution can be well defined and developed and lead time can be reduced, along with work-in-process and labor costs. In this chapter, line-balancing techniques will be explained and developed.

THE ASSEMBLY LINE

The assembly industry has evolved since appearance of the first assembly line during Henry Ford's time. The first movable assembly line was created by Ford to manufacture the Ford Model T. Ford installed a capstan and a thick cable (Fig. 3.1) to move the cars between assembly stations.

Before this development, cars were manufactured in a fixed-position layout, where materials and workers had to be moved to the car's location. Because the new "drag line" pulled the vehicles to various workstations, the workers stayed in one place, and the cars changed locations for the appropriate assembly tasks.

With this new assembly line, the time needed to manufacture a car was reduced from 13 to 6 hours. The original assembly line produced only identical black Ford Model T cars because it was the color that dried the fastest, decreasing the car's price and thereby allowing many Americans to afford a new car.

Ford used the specialization ideas developed by Adam Smith. According to Smith, in the manufacture of nails, it was better having one worker making the head and another worker making the tip instead of having a single worker conducting both operations.

Specialization and task decomposition lead to the birth of such new professions as industrial engineering. Early on, industrial engineers were devoted to component task design and to manufacturing scheduling, defining the best way to perform various production activities.

Later on, Alfred Sloan, at General Motors, enhanced Ford's work, making it possible to mix different models on the same assembly line. During the 1960s and 1970s, Japan did not adopt the Ford, Sloan, and Taylorist way of working. Instead, the Japanese expanded worker involvement so that workers were able to handle a variety of manufacturing processes. This fact was a key for development of the cellular layout.

THEORETICAL BASIS

Mass Production

A broad range of products is manufactured in very large batches so as to satisfy mass market demand. Some of them, such as food products, have very specific manufacturing requirements and processes and will not be addressed in this chapter. Other products, such as toys, cars, and electrical appliances, which share similar manufacturing methods, will be studied in depth.

Mass production has two basic characteristics: low prices compared with the cost of the handmade products and the ability to replace component parts on the manufactured models that break or wear out.

Figure 3.1. Ford's assembly-line concept.

- The competition for lower prices mandates a decrease in manufacturing costs and supposes that the development of a product should be as standard as possible. The flexibility of the assembly-line dictates the variability of the family of products that can be manufactured.
- Reducing the number of different replacement parts (standard replacements for a variety of original equipment components) allows a significant reduction in the number of spare parts inventoried for different models, simplifying maintenance tasks. It is important to remember that companies must maintain a spare parts accessibility for a certain amount of years after the product is withdrawn from the market.

On the one hand, simple product assemblies reduce process time, allowing firms to manufacture more articles in the same or less amount of time. As a result, the cost of each article decreases. On the other hand, if the complexity of a product can be increased to accommodate additional replacement part applications, then the part's utility increases.

A product-quantity (P-Q) analysis tool helps, using Pareto analysis, to decide the optimal plant distribution for the different products manufactured. When the product quantity is large and its variety is small, massive production fits perfectly, and focused (not particularly flexible) assembly lines provide the most economic production alternative.

Flow or Assembly Lines

The recommended way to manufacture articles that are mass produced is by using flow or assembly lines. Flow lines are used to produce individual components, and then assembly lines are used to assemble the components into worthwhile products. These systems consist on workstations dedicated to the progressive manufacture and assembly of parts and are integrated using materials handling devices that allow parts to move from machine to machine or workstation to workstation. Creating flexible cells generally requires the duplication of machines, and as a consequence, the production system cost increases. In order to avoid this problem when manufacturing large volumes of standard products, it is recommendable to invest in several simple machines with low cost instead of investing in a versatile and more expensive machine (Fig. 3.2).

Another advantage of assembly lines is that the work-in-process decreases because containers with waiting parts are placed in small queues (often one) in front of the machine. This design eliminates excessive inventory typically required when using more versatile and expensive machines. The inventory is required because it is more critical to have higher machine utilization of more costly equipment, and it is more difficult to manage a greater variety of incoming material.

The most frequently used materials handling device is the conveyor belt. Depending on the size of the product, the conveyor belt might be replaced by overhead conveyor systems, where chains are used to move and support a hanging product.

This method often increases the effective space utilization of a factory and can reduce materials handling time significantly. For overhead systems, each product either (1) stops a predetermined amount of time at each station, called *cycle time,* a basic concept in line balancing that will be studied later on (this technique is called *synchronized flow*), or (2) is removed at a workstation to be worked on (this technique is called *asynchronous flow*).

An effective layout when considering product movement, multifunctional workers, and one-piece flow is called an *island layout* (Fig. 3.3). However, this layout has the disadvantage that the worker is specialized in an island or subject and therefore not multifunctional. Also, the islands are isolated with respect to each other, and synchronization among them can be difficult.

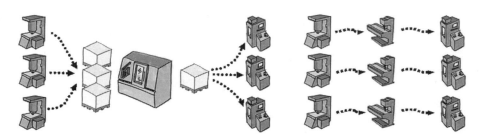

Figure 3.2. Versatile machine versus specialized machines.

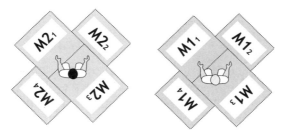

Figure 3.3. Island layouts.

Cell Layout Design Justification

The manufacturing environment has changed significantly in the past several years. The requirement to meet specific customer needs has created the need for mass customization of products, where various product attributes can be specifically tailored to a particular customer. For instance, today it is possible to order jeans that are specifically sized and designed to fit an individual.

At the same time, it is necessary to get the custom product to the customer in as short an amount of time as possible—otherwise, a competitor with a shorter lead time will earn the business. Table 3.1 summarizes some of the trends among manufacturers. All the trends noted in the table point to the increased use of more flexible manufacturing cells.

The focus of flow or assembly lines has been on the use of process-focused systems or job shops. These systems maximize flexibility but have other problems. For instance, other reasons to suggest the use of the manufacturing cells as opposed to process layout include the following:

- A high number of indirect workers (materials handling or maintenance workforce) is generally required in process layout.
- A high level of work-in-process is also typical of process layout.

TABLE 3.1. Previous and Current Economic Environments

	Previous	Current
Product	Little variety	Great variety
Delivery time	Not very demanding	Very demanding
Size of lots	Large	Small
Lead time	Long	Short
Product life cycle	Long	Short

- Product quality responsibilities are not as clear when using process layout.

When a cellular layout is employed properly, unnecessary product movement can be eliminated, product and personnel flexibility can be achieved, and a more economic production environment can be obtained. However, the transition to cells is not easy, as will be shown in this chapter.

Basic Cell Design Nomenclature

Task. Each *task* corresponds to the necessary steps that the work gets decomposed into for product transformation. A task is considered the smallest assignable unit to a cell, and it is critical to define the beginning and the end of a task in a precise way.

Methods and time methodologies facilitate the goal of optimizing a task and will be explained in more detail in Chap. 5. It is also necessary to know the task time. Time study is also briefly presented later in this chapter so that standard task times can be determined.

Workstations. *Workstations* are the specific manufacturing or assembly stations defined to perform task(s). The number of workstations needed must be as small as possible so that a minimum workforce can be maintained along with a reduced work-in-process.

Takt Time and Cycle Time. *Takt* is the German word for "rhythm" or "pace" and is a critical term for manufacturing systems design. *Takt time* is the allowable time to produce one product at the rate a customer demands it. This is *not* the same as *cycle time,* which is the normal time to complete an operation on a product in each workstation (which should be less than or equal to *takt* time).

Takt time is defined as the result obtained from dividing the useful working time in one day by the daily customer demand. Cycle time is the sum of the task times that a product requires at each station.

Logically, during the cell design project, it is important that the *takt* time and the cycle time be as close as possible. However, because of imbalances in workstation activities, each workstation cycle time may in fact vary. Also, owing to the fact that the *takt* time varies with respect to the real demand, they will not always coincide.

Takt time never can be smaller than the largest workstation cycle time; otherwise, demand rhythm becomes faster than the production system is capable of handling.

Knowing the demand that should be satisfied by the line or cell, the maximum cycle time allowed at each workstation to perform the task can be calculated:

$$Takt \text{ time} = T_c = \frac{\text{useful working time in one day}}{\text{daily demand}}$$

Cycle time is a measure of how much time it takes for a particular operation, which is also expressed in similar units (time/piece). *Takt* and cycle times are illustrated in Fig. 3.4.

In the figure, the upper illustration shows an incorrect implementation of *takt* time where an average cycle time (or just below the average cycle time is used as *takt* time). In this scenario, operator C will not be able to keep pace with the rhythm of the system, thereby creating a "bottleneck." In the lower illustration, all the operators will be able to keep pace with the system rhythm or *takt* time. The *takt* time might be determined for short periods by the maximum cycle time at one of the workstations in the process (the bottleneck).

This production control discussion is a much-abbreviated discussion of control in a lean system. It is intended as an overview of how an effective lean control system works. When the workstation with the

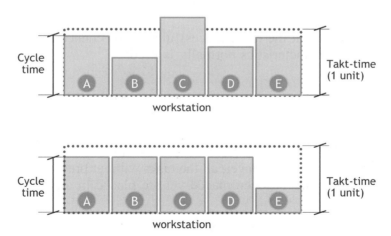

Figure 3.4. An illustration of *takt* time and cycle time comparison.

largest cycle time has finished, if the line has a conveyor belt, it will move the product to the next workstation. Therefore, all the workstations must have finished the assigned tasks within this time.

Total Workstation Cycle Time (p_i). Generally, more than one task is assigned to each workstation. The sum of the process times of each one of the tasks determines the total cycle time assigned to the workstation.

Idle Time (h_i). The workstation idle time corresponds to the difference between the *takt* time and the cycle time, or total work at each workstation. The idle time never can be less than zero because it would suppose that a workstation has assigned more work than the *takt* time. As a consequence, when the line moves, the worker is not going to conclude all the assigned tasks.

Reality for many assembly companies can be quite different from theory. Cell flexibility allows a company to rebalance tasks on the occasions when the maximum cycle time exceeds the required *takt* time in the assembly of a specific product. Therefore, the problem is scheduling batches of product to the line to achieve a feasible mix of products.

Precedence Diagram. The assembly of a product is not a random activity; typically, it requires a sequence precedence. It is necessary to know the operations summary and precedence restrictions that exist when assembling the product.

In order to assign the tasks successfully, a precedence diagram with the following characteristics normally is used:

- The tasks are arranged so that there are only left-to-right arrows, and there are no precedence relationships between tasks in the same level.
- All the tasks without any precedence are placed in the first level.
- In determination of a level, all the tasks without precedence among those that have not been placed yet are placed in that level.

The figure also illustrates that for each box of the diagram, next to the task name, the task duration is included (Fig. 3.5).

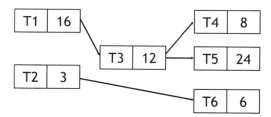

Figure 3.5. Example of a precedence diagram.

CELL DESIGN METHODOLOGY

The methodology to develop production cells is very straightforward. The different steps that should be carried out are as follows:

- Form product families.
- Change the locations of machines.
- Calculate the output production rate, the task assignment in each workstation, and the necessary number of workers in the cell.
- Plan and control the cell.

However, the difficulty lies in the high number of prerequisites needed to be able to carry out this layout transformation:

- Multifunctional workers
- Determination of the required space for the cells (specially U-shaped cells)
- Investment in new machinery
- Improvement in the setup time of dies
- Looking for simple methods for production automatization
- Choosing new production planning and control systems so that future planning or control of the line is not needed

CELL DESIGN TOOLS

Line-Balancing

The main goal of line-balancing techniques is to assign tasks to workstations so that the minimum number of workstations can be achieved, where each task needed to produce a part is assigned to only one workstation. It is important to take into account that in each workstation the assigned work p_i is less than the *takt* time, and the idle time h_i is minimal.

This assignment of tasks to workstations should not violate any of the precedence relationships among tasks imposed by the product design and production method. These relationships are represented using a precedence diagram, as shown in Fig. 3.5.

Line-balancing for some companies may require specific algorithms, where the assignment of tasks is made in a different way. Instead of creating workstations according to the task assignment, the workers should be assigned to a fixed number of workstations. In these cases, the next methodology does not apply.

General Steps in the Line-Balancing Methodology. The process of line-balancing always follows the same outline:

Step 1. Define the tasks and their times (t_i).

Step 2. Specify the precedence relationships, building the precedence diagram explained previously in this chapter.

Step 3. Determine the *takt* time T_c also explained previously in this chapter.

Step 4. Calculate the minimum number of workstations M_{min}. It is necessary to round the number up. M_{min} supposes the lower limit of the number of workstations that can be created:

$$M_{min} = \left| \frac{\sum t_i}{T_c} \right|^+$$

Step 5. Choose a tasks assignment rule. (This is explained later on.)

Step 6. Assign tasks until the assigned time is equal to the *takt* time or until it is no longer possible to assign a task owing to the restrictions of time or sequence (the task time plus the assigned time exceeds the *takt* time). In this case it will be necessary to create a new workstation and continue with the assignment.

Step 7. Determine the total idle time and the line efficiency:

$$H = M \cdot T_c - \sum_{i=1}^{M} p_i \qquad \text{Efficiency} = \frac{\sum_{i=1}^{M} p_i}{M \cdot T_c}$$

If the idle time of each workstation is 0, the line is perfectly balanced.

Step 8. If the solution obtained is not considered acceptable, it is necessary to choose another assignment rule.

Tasks Assignment Rules. A task is eligible if it has not yet been assigned and all those that precede the task (preceding tasks) have been. Frequently, it will be necessary to decide among several eligible tasks. There are no optimal methods to select the task that will be assigned to the workstation, but there are a number of heuristic methods that lead, in many cases, to an only and almost optimal solution.

Heuristic methods are simple rules that propose two selection criteria. The second criterion will be used only if there are several tasks that coincide in the first criterion.

The solutions obtained with application of the various heuristic rules can be compared by analyzing the idle time share among the workstations, studying the value of the efficiency and checking if the heuristic selects the minimum number of workstations M_{min}.

The following subsections present the most used heuristics and the decision criteria used in each of them.

Total-Number-of-Following-Tasks Heuristic

- Among the eligible tasks, choose the task that has the largest total number of following tasks.
- If two or more tasks coincide in the first criterion, select the task with the longest time t_i.

This method tries to move through the precedence diagram as quickly as possible.

Individual-Durations Heuristic

- Among the eligible tasks, choose the task with the longest time t_i.
- If two or more tasks coincide in the first criterion, select the task that has the largest total number of following tasks.

Largest-Positional-Weight Heuristic

- Among the eligible tasks, choose the task that has the largest positional weight. Positional weight is the sum of the task time and the time of all its following tasks.

- If two or more tasks coincide in the first criterion, select the task with the longest time t_i.

Line-Balancing Special Cases

Task Time Larger than the Takt Time. It has been mentioned already that in some cases (mainly in those with automatic processes carried out by machines) it is possible that the task time may exceed the *takt* time. In such cases, there are two possible choices:

- Accept that the task dictates the *takt* time.
- Solve the problem, for example, by decomposing the specific process into more than a single operation.

The first choice supposes the loss of possible product sales or an increase in the number of pending orders because before this choice the *takt* time was fixed by the demand, and therefore, the production rate was equal to the demand rate. When restricting *takt* time to a particular task time, production will be reduced and will not be able to satisfy all the orders.

To solve the problem of a task time larger than the *takt* time, there are some alternatives to consider:

- *Divide the task into two tasks.* If the task is carried out manually, it could be possible to separate it into two tasks. It will be necessary to analyze and define the task again.
- *Improve the task or the product.* This can be done by workforce optimization, which will be explained at the end of this chapter.
- *In the case of an assembly task, incorporate an assistant.* This is possible whenever the part size allows it.
- *Place two workstations in parallel.* This solution can be observed in the Gantt diagram in the Fig. 3.6. It increases the work-in-process, and the products are assigned to each parallel workstation every two cycles. This solution will be adopted if and only if the other choices are not possible.

Line Balancing in U-Shaped Cells. The line-balancing method sometimes causes an unequal time assignment. The U-shaped layout with shared tasks helps to solve these unequal-times assignment situations, as shown in Fig. 3.7.

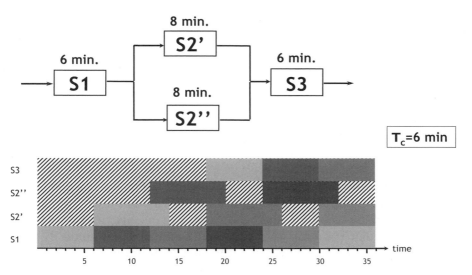

Figure 3.6. Two workstations in parallel solve the problem of a task time larger than the *takt* time.

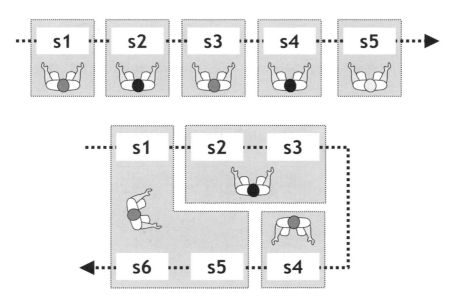

Figure 3.7. Line balancing in a U-shaped layout.

U-shaped cells (Fig. 3.8) with entry and exit points located near each other avoid constant displacements to the start of the line at the beginning of a new part and solve many of the island distribution problems (Fig. 3.3). The worker moves along the cell, and the movements are minimized.

The U-shaped layout improves task assignment by offering production rate flexibility. In other words, the number of workers assigned can be changed at any time. Then it is possible to assign different numbers of workers, which makes it easy to adapt the cycle time to the *tack* time without rearranging task assignments.

In addition, the U-shaped cell is the most flexible layout when facing changes in the number of workers owing to demand fluctuations. The main advantage of this layout is the flexibility because it is possible to vary the production output by incorporating new workers in the system, as shown in Fig. 3.9. The production line in this figure can produce parts with very different rates depending on the number of workers assigned to the cell.

The mathematical resolution of U-shaped line balancing needs to be solved in a good way. However, solutions can be obtained in two different ways:

- By means of observation of the workstations and their corresponding idle times, as happens in the Fig. 3.7. This is not always easy.
- By making the line balancing suppose a *takt* time $T_{c'}$ equal to half the *takt* time ($T_{c'} = T_c/2$) and later by bending the line (Fig. 3.10). A problem arises, however, when the number of workstations is odd.

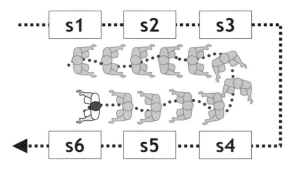

Figure 3.8. U-shaped cell.

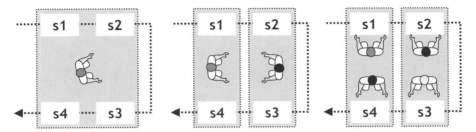

Figure 3.9. Flexibility in the production output in a U-shaped cell.

U-shaped Line Balancing with a Task Time Larger than $T_{c'}$. The U-shaped line-balancing problem should analyzed the problem of having a task with a larger time than $T_{c'}$. In the case that the task time is larger than $T_{c'}$ but smaller than the *takt* time T_c, the task will be assigned to an independent workstation, as shown in the Fig. 3.11. It is possible to assign other tasks to complete the cycle time for the workstation.

In the case that the task time is also larger than the *takt* time, it will be necessary to place another workstation in parallel (Fig. 3.12).

In both these cases, the line looses flexibility but retains other benefits, such as the rotation in working positions or better materials handling.

Group Technology

The formation of self-managed subfactories is not always simple. For example, which products are grouped to create a cell? How many cells

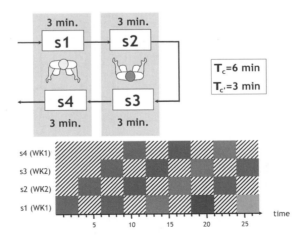

Figure 3.10. U-shaped line balancing done with a *takt* time equal to $T_c/2$.

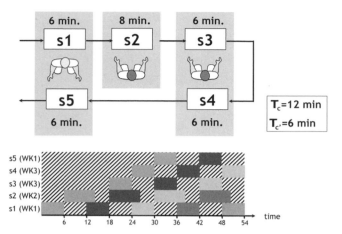

Figure 3.11. U-shaped line balancing with a task time larger than $T_{c'}$.

should there be? There are specific methodologies, specifically those that have been used in group technology, to facilitate the formation of product families.

Group technology is a tool used in engineering and production departments to identify similar products and to group them into product and/or production families. As a result, economies of scale are achieved, forming multiproduct families rather than from a single part. By using the concept of group technology, many small companies can

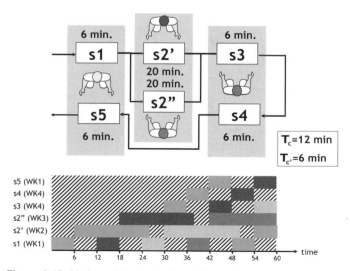

Figure 3.12. U-shaped line balancing with a task time larger than T_c.

compete against bigger companies by offering seemingly unique products but with numerous similar attributes.

Some automotive companies use the same platform to manufacture car models that on the surface appear to be quite different from each other. The company may use the same drive train and suspension for vehicles with different purposes. Despite the fact that the vehicles look different and serve different uses, some standard components still may be used. This concept is illustrated in Fig. 3.13.

In some cases, alliances have taken place between different companies to manufacture new models sharing the development costs. In conclusion, the goal of group technology is to reduce design and production costs.

Many of the gains of group technology come from design standardization. Not only are standard parts used, but also in some cases the components are designed in a way that allows them to be used in different products; for example, some motorcycles use the same screw for different fastening needs. This screw is gauged according to the most critical point, but the company saves money by extending the use of the same screw in other motorcycles because inventory management and purchase costs are reduced.

Family Formation. It is usually impractical to have a manufacturing cell for each product because the necessary investments would be bigger than the expected benefits. Therefore, grouping products into larger families eliminates this problem by decreasing capital product investments.

In some cases, the formation of families can suggest small design changes in a new product that, if the changes do not affect the product functionality or to the client's needs, allow the product to be produced in the same manufacturing cell as the rest of the family.

There are several techniques to form product families—from the most basic, based on direct visual inspection and classification, carried

Figure 3.13. Different types of vehicles still can use standard drive trains.

out by some expert, all the way up to complex mathematical methods of product features and shape.

Advantages of Group Technology. The savings obtained from group technology applications are numerous, and some of them are even quantifiable. For instance, savings of up to 50 percent have been seen for the design time, avoiding the duplication of designs, as well as saving for work-in-process of as much as 60 percent and reductions in time-to-market of 70 percent. These benefits in production come from different sources:

- Reduction in the number of dies required for press working
- Reduction in purchasing cost by clustering and grouping orders of similar components
- Scheduling simplification owing to a reduction in the number of product types and the formation of self-managed cells

Disadvantages of Group Technology. It would seem that seeing the numerous advantages that group technology offers, the applications of group technology should be extended. Although it is true that every day more companies apply group technology, there are some disadvantages that impose barriers to the adoption of this method:

- The family classification consumes time and effort and does not always lead to profitable solutions. It also can suggest erroneous changes in a product's design.
- The installation of manufacturing cells demands changes in work habits because the responsibility is distributed among workers. In addition, layout changes usually require large investments.
- The duplication of some machines, in order to create cells, implies a low utilization of equipment that may have been better used before the change. Traditional business management techniques, based on productivity, penalize these reductions in production rates.

Time Study

Time study can be defined as the methodology used to determine the standard time for an operation—in other words, the time that a qualified worker needs to carry out a specific task, working in a normal rhythm during a work day. The first historical work in the establishment of operation times go back to Fredrick Taylor. Taylor used the most qualified worker to establish the standard working time. Taylor's work consisted on educating the rest of workers to carry out the tasks in the same manner with the same rhythm.

Taylor was opposed by labor unions, which said that Taylor's method had killed a worker. This was not true, but after the incident, Taylor decided to include two kinds of allowances: those related to the worker fatigue and those involving variations in worker capabilities.

Frank and Lillian Gilbreth added to the body of knowledge started by Taylor. In their analysis, they divided tasks into fundamental elements and chose the best worker for each task so that the standard time was obtained by adding the best times of each task. Years later, Lowry, Hayrard, and Stegemerten, at Westinghouse company, defined a normal-worker concept that is still used today.

Time studies lead to establishment of the *work standard,* which is considered to be another tools in the just-in-time (JIT) philosophy (Fig. 3.14).

Therefore, standard time establishment has other important utilities:

- The use of standard times for factory scheduling
- Providing standards for incentives systems if the system is based on worker productivity
- Comparing different work methods
- Optimizing the number of workers required to achieve a schedule
- Knowing the production costs

There are four different methods to determine the standard time of an operation:

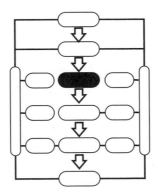

Figure 3.14. Location of work standard in just-in-time schema.

- Time study with chronometer or videotape cameras
- Normalized times
- Predetermined time
- Work sampling

The last two techniques will not be addressed in this book because the first two methods are used the most, and they lead to good solutions.

Time Study with a Chronometer or Videotape Cameras. This methodology allows an industrial engineer to obtain the standard time for an operation by means of direct task observation. Registration of the time is carried out with a stop watch (chronometer) or a camcorder. This activity is illustrated in Fig. 3.15.

Study Preparation. Before recording times with a chronometer, it is necessary to clarify that the goal is to determine the standard time, not the time that the worker is really using during the observation. Workers typically speed up the process when they are being watched. Therefore, the time obtained should be one that a worker would use without knowledge that the task and methods are being assessed.

Once this standard time is set, it can be compared with the time typically required to carry out the operation. The deviation between the times can be used for future improvements.

Before carrying out the study, it is necessary to become familiarized with the process or task being studied and the part that is processed. It is also recommended that a layout of the working environment be developed so that worker movements can be mapped while the worker is being observed.

Figure 3.15. Time study with a chronometer is similar to sports coaching.

It is also important to determine the work cycle being studied in order to understand the time and working process. For example, cycle time can be the time from the inspection of a product until the inspection of the next product.

Time study is not only a technical problem of measuring the time required by a person to carry out an operation because human behavior enters the process, and the operator can speed up or slow down. Understanding and assessing the operator are basic requirements for a successful time study. According to some experts, technical knowledge is only 25 percent of the requirements for a time analyst. The analyst should formulate questions in a polite manner and inform workers that the study will be used as a reference. It is a matter of vital importance to convince workers to contribute to the study, as well as to request their suggestions. Of course, workers neither should be criticized nor should they be corrected while the study is being carried out. Such actions usually lead to an erroneous study.

Data Acquisition. The chronometers used in the time measurement can be special, although the standard method is to use a traditional stop watch. Special chronometers (called *minute decimal chronometers*) have a 1-minute cycle with 100 divisions. As a result, by adding the timed hundredths, the result is obtained in minutes.

A time study should be broken into working elements. In other words, times for the basic movements should be measured. It is also important to determine the beginning and the end of each element. During the study, it is necessary to distinguish the following element categories:

- *Repetitive and casual (nonrepetitive) elements.* If, for example, a machine is greased after each 100 parts produced, the greasing should be considered as a casual element and should be considered separately from the cycle time.
- *Constant and variable elements.* For variable elements (activities whose duration will depend on a specific task), the time of execution depends on a product or equipment characteristic (e.g., the part weight) or the type of equipment. In this way, the unnecessary use of time is avoided.
- *Identify accidental elements.* These elements (e.g., dropping a tool) are not part of the study and should not be included in the standard time.
- *Separate machine elements from worker elements.* Taking into account the separation of resource dependencies is important. For

example, a worker may have to wait for a machine to finish its cycle before he or she can remove the part. The worker should not be penalized when the other resource is busy.

- *Separate unnecessary and essential movements.* An inexperienced worker may execute unnecessary movements.
- *Identify simultaneous movements.* The longest time within the durations of the simultaneous movements must be considered as the time of the combined movements.

Techniques for Data Acquisition. There are basically two different methodologies for data acquisition:

- *Snapback method.* The chronometer is stopped and returned to zero at the end of each task. Sometimes it is not possible to register all the activities in each cycle because of the time needed to return the chronometer to zero.
- *Continuous timing.* The chronometer does not stop until the end of the study. The analyst reads the time in each task, but the time is presented in an accumulate way. After the study, the elemental task duration is calculated by subtracting one time from the following one. This procedure is better in practice, than the preceding one mainly if the videotape is used.

Determining the Number of Cycles to Study. In any time study, there will be variations among the times needed to do the same activity. The needed number of cycles to perform an accurate time study depends on the duration of the elements as well as on the accuracy required. There are numerous ways to determine the needed number of cycles n, and most of them depend on the data obtained in a first step of the study:

$$ n = \left[40 \frac{\sqrt{n' \sum x^2 - \left(\sum x \right)^2}}{\sum x} \right]^2 $$

where n' is the preliminary study size (in a first moment, usually 10 readings are taken), and x is the amount of time measured. This expression considers the time study as a sampling procedure in statistical terms. The needed measures are the difference between n and the number of measurements taken to calculate the expression n'.

Standard Time Establishment. Once the times for *n* cycles have been measured and recorded, it is possible to proceed in the standard time establishment. There is an established procedure that facilitates this task, and it will be explained next.

The first stage is to eliminate the measurements that result from accidents, mistakes, etc. The other measurements can be considered significant. The average of these observations gives the first time of the task, called *observed time OT,* but it should be corrected by taking into account two factors:

- The speed at which the worker did the task when the study was performed
- The possibility of maintaining the same working speed during the whole day

Normal Time NT. A *normal speed* is understood as the time that a normal worker can reach and maintain during a theoretical working day without additional fatigue (Fig. 3.16).

Companies usually demand, compared with normal, a higher working speed denominated rhythm type or activity using three time scales (Table 3.2):

There are numerous factors that must be taken into account when the normal time is being calculated; some of these factors do not depend on the worker, such as changes in material quality, tool wear, etc. However, other factors are exclusive to the worker and his or her working speed.

The worker's attitude regarding organization or professional know-how determines the worker's working speed. The analyst should fix a subjective value to the observed working speed. This correction is called the *activity factor AF*.

Figure 3.16. Normal time does not consider worker fatigue.

TABLE 3.2.   Most Common Activity Scales

Normal speed	Rhythm type or activity
100	133
60	80
75	100

$$NT = \frac{\text{observed speed}}{\text{rhythm speed or activity}} \quad OT = AF \cdot OT$$

Allowances A. The allowances for personal necessities (resting and recovering from exhausting tasks) vary between 5 and 7 percent of *NT*. The allowances for basic fatigue (weigh lifting, environmental conditions) use to be a 4 percent.

Other components concern unforeseen events or special process characteristics (if the process itself produces idle time). All these allowances increase *NT* and lead to the establishment of a *standard time* (Fig. 3.17).

Given the complexity of determining the value of all the suitable allowances, a fixed factor *A* that oscillates between 13 and 15 percent is used frequently. That is,

$$ST = A \cdot NT$$

As a result of this time study, the *standard time* is defined as the time needed for a well-trained worker to carry out a task using an

Figure 3.17. Personal necessities.

established method and working to a normal speed during a real working day.

Predetermined Time. Sometimes the analyst has to determine the standard time of an operation that does not yet exist, e.g., a new process that is being designed. In this case, the chronometer cannot be used. If the operation can be broken down into a limited number of elementary tasks, and if the duration of these tasks could be determined, it would be possible to set the standard time of the operation in a simple way.

Everything in life has limitations, and worker movements are no exception. As a good analogy, the English alphabet is formed with 26 letters, and any text can be broken down into 26 different letters. Similarly, there are a limited number of basic movements for workers (Fig. 3.18). The analysis and study of the determination of those basic movements and their lengths are carried out by predetermined time systems.

These predetermined systems have the advantage of leading to more coherent standard times. However, they are very difficult to implement and understand, and the use of working tables provided by these methods can be very complex. Another important disadvantage that should be considered is the fact that these systems do not eliminate the need to use the chronometer because the machine manufacturing times and waiting times are not included in the tables.

MTM System (*Method-Time Measurement*). The *method-time measurement* (*MTM*) *system* is one of the most extended predetermined time systems. This method covers generic movements of a person, e.g., release, reach, move, turn and apply pressure, grasp, position, and disengage.

Figure 3.18. Predetermined time considers elementary human movements.

Three additional movements are also included: two eye movements (travel and focus) and an upper body movement. Lastly, the system considers 13 movements that belong to the body, and all the movements are tabulated.

The combined movements are also tabulated, that is to say, those that take place when two or more movements are executed with the same body part simultaneously. The same thing can be said of the simultaneous motions that correspond to movements carried out by different body parts. The MTM system includes a table that shows the difficulty of the different combinations of basic movements. The time units that the MTM system uses are the *time-measurement unit* (TMU). 1 TMU = 1/100.000 hours = 0.036 second.

Simplified MTM System (MTM-2). The International MTM Directorate developed a simplified method called *MTM-2*. This method leads to more accurate solutions, with an error of only 5 percent, becoming more efficient than the traditional MTM method.

Leveling Production

At the beginning of this chapter, the prerequisites for conducting a cell layout transformation were presented. One of these prerequisite activities occurred at the company operation level, i.e., production planning and scheduling strategy. Cell layout transformation requires a different set of tools to carry out these planning and control processes.

A planning tool used in the JIT philosophy is called *leveling production,* and it is included in both the JIT and the 20 keys (key number 16) for lean manufacturing (Fig. 3.19).

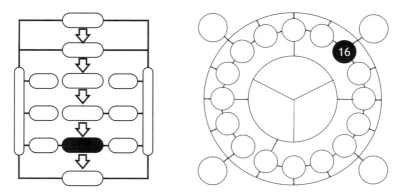

Figure 3.19. Leveling production location in JIT and 20 keys diagrams.

To explain the concept of leveling production, Tomo Sugiyama used a military example, namely, the battle of Nagashino. In the battle of Nagashino, Nobunaga opposed Takeda. Nobunaga headed a thousand soldiers with muskets (a new weapon at that time). Takeda commanded a traditional army cavalry. Takeda hoped to be able to attack Nogunaga while the muskets were being loaded (Fig. 3.20). However, he was not able to get in while the muskets were being loaded, and Takeda lost the battle. Why?

Nobugana's victory was due to a strategy that he devised. He divided his army into three parts so that there was continuous firing. This strategy allowed a third of his troops to be ready at all times (Fig. 3.21). As expected, a third of the soldiers were always firing at Takeda's army. Therefore, Takeda did not find the right moment to attack.

Nobunaga's approach also can be applied to factory production, as will be explained. In mass-production systems, it is possible to calculate the *takt* time, as explained earlier in this chapter. In this type of factory, each product is manufactured in an independent production line (Fig. 3.22). Therefore, production scheduling is not affected only by product leveling but also by fulfilling customer due dates.

In cellular layouts, several products typically share the same cell as well as schedule. Thus manufacturing different part families and product families becomes more complex.

The initial strategy for most companies is to carry out traditional constraint-based monthly planning to satisfy customer demand (Fig. 3.23). As a consequence, a large amount of inventory is produced, maintained, and stored.

If a daily planning is conducted instead of monthly (Fig. 3.24), the amount of inventory generated would be 20 times smaller (supposing 20 working days).

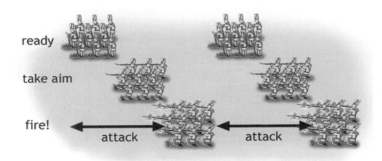

Figure 3.20. During musket loading time, the soldiers were undefended.

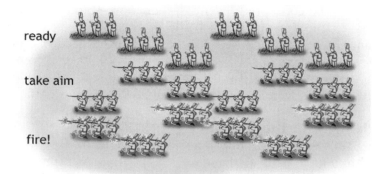

Figure 3.21. Nobunaga's strategy: a third of the soldiers firing constantly.

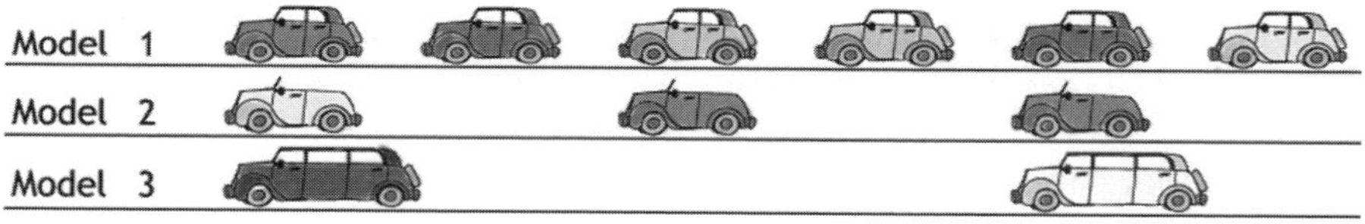

Model 1
Model 2
Model 3

Figure 3.22. In mass-production systems, scheduling depends on customer requirements.

480 u/month
320 u/month
160 u/month

Figure 3.23. Monthly planning strategy leads to large inventories.

24 u/day
16 u/day
8 u/day

Figure 3.24. A daily planning strategy partitions inventories into smaller temporal states.

Leveling production, according to Nobunaga's approach, leads to "constantly firing" of different products. This tool forces a company on changing its production scheduling strategy. Leveling production calculates the *takt* time for each product, as shown in Fig. 3.25.

Nevertheless, the development of a feasible sequence that satisfies demand is quite complex. In the preceding example, it is possible to find a proper sequence (Fig. 3.26). If this sequence is repeated 10 times, it fulfills the demand presented in Fig. 3.23.

Generally, without applying mathematical methods, it is difficult to obtain a feasible (or optimal) solution when manufacturing different products. In order to achieve this goal, production planning and scheduling in the factory should be leveled, which means that the needed production steps must be synchronized to favor a continuous one-piece flow. Obviously, if cell layout has not been implemented properly, level production is not possible.

As in most production systems, most of the components used in a car manufacturing process are not processed in the same plant. As a result, leveling production forces the suppliers to level their manufacturing processes; otherwise, they will need to store large quantities of finished products to satisfy the demand.

This manufacturing strategy by assembly plants has created one of the main obstacles to implementation of the JIT philosophy. Some component suppliers have implemented draw-from inventory systems, where they store their inventory at the assembly plant. The supplier owns the inventory until the assembly plant draws items for assembly. This procedure requires that the assembly plant has the necessary storage capacity to house the inventory.

Again, the difficulty lies in the large number of prerequisites needed to implement production leveling:

- Repetitive production
- Excess capacity in productive resources
- Fixed production for, approximate, a time horizon of one month

Figure 3.25. *Takt* time of the production of each family manufactured in the cell.

Figure 3.26. This sequence fulfills the demand presented in the example.

- High inventory costs
- Low resource costs
- Multifunctional workers
- Short setup times

Many of these prerequisites are included in the cell layout prerequisites; therefore, if the cell is designed properly, leveling production is possible.

Multifunctional Workers

Until the 1960s, the standard was for each machine to have a dedicated worker. Specialization of workers to specific tasks was the main productivity concern for a company.

As market needs have changed from stable long-life products to the highly dynamic products of today, the requirements for production workers also have changed. Today's market demands workers who are familiar with more than one process. Production workers need to carry out operations that may involve multiple departments and even quality and maintenance tasks. Multifunctional training has become a standard for today's factory worker (Fig. 3.27).

Multifunctional workers are an important key to the JIT philosophy, as well as the 20 keys (key number 15) for lean manufacturing (Fig. 3.28).

Figure 3.27. Specialized versus multifunctional worker.

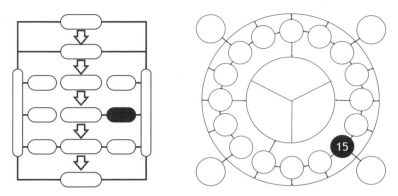

Figure 3.28. Multifunctional worker locations in JIT and 20 keys diagrams.

Task rotation is a good strategy to maintain the multifunctionality of workers. It has multiple advantages, such as reduction in work-related accidents, improved relationships among workers, and enhanced knowledge sharing.

Task rotation also helps to uncover expert workers in each task. These workers should instruct other, less skilled workers on the specifics of a particular activity.

Training programs can be used to create comfortable learning environments and to promote the self-improvement of other employees. It is worthwhile to study the possibility of providing incentives to workers who show an interest in learning and sharing their knowledge of other machines with other workers (Fig. 3.29).

The time and effort invested in training programs always suppose a reduction in the company's production capacity; however, the ultimate

Figure 3.29. Training programs are vital to generate multifunctional workers.

goal for training is to have a positive long-term effect on production capacity.

Workforce Optimization

Workforce optimization is a tool that is part of the JIT philosophy. The goal of this tool is to define how many people are necessary to make what it is necessary to make; i.e., its objective is to manufacture the product(s) with the minimum number of workers (Fig. 3.30).

One of the principles of optimization is not to assign a fixed number of workers to a line (Fig. 3.31). Thanks to multifunctional workers, it is possible to use the available workforce in the best possible way.

For example, supposing a cell formed by five workstations with the task assignments as shown in Fig. 3.32, the tasks can be redistributed so that one workstation can be eliminated. Once the first objective in optimization of the workforce is achieved, it is time to reduce the workforce by means of better task assignment.

After redistributing the tasks, it is important to improve the task by eliminating a second workstation instead of distributing the work among the four workstations (Fig. 3.33).

The time needed to carry out a task can be decreased by eliminating waste and operations that do not add any value to the product (Fig. 3.34). This is the way to eliminate the need for a fourth worker in this cell. In order to eliminate waste, there are method-improvement tools that will be analyzed in Chapter 5.

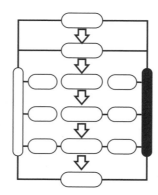

Figure 3.30. Location of workforce optimization in JIT schema.

Figure 3.31. Fixed assignment of workers.

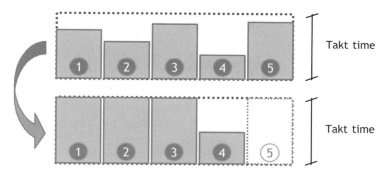

Figure 3.32. Redistribution of tasks in a cell.

Figure 3.33. Workforce optimization tries to eliminate another workstation.

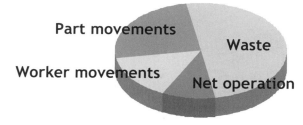

Figure 3.34. Waste and operations that do not add value to the product.

SUMMARY

In this chapter the methodology and requirements for the design of cellular layouts have been developed. U-shaped layout has been chosen as the best cell design, although this layout requires multifunctional workers in order to be efficient. In addition, this chapter has presented some methodologies, such as time study and group technology, that must be used to effectively carry out cellular design. Nevertheless, use of these methodologies is not limited to the design of manufacturing cells; they can be used in other improvement areas as well.

RECOMMENDED READINGS

José Manuel Arenas, *Control de tiempos y productividad*. Paraninfo, Madrid, Spain, 2000.

Luis Cuatrecasas, *Diseño de procesos de producción flexible*. Productivity Press, Barcelona, 1996.

Benjamin W. Niebel and Andris Freivalds, *Methods, Standards and Work*. 11th ed. New York: McGraw-Hill, 2003.

4

Equipment Efficiency: Quality and Poka-Yoke

Shigeo Shingo developed a system to improve inspection tasks with the goal of guarantying 100 percent quality for manufactured parts, leading toward a defects-free process. This chapter will explain inspection processes based on unnoticed mistake-proving devices (called *poka-yoke*). This type of inspection strategy complements statistical process control (SPC) and is used primarily for inspecting logical features.

Poka-yokes are visual and physical tools that should be used in conjunction with source inspection (a concept also created by Shingo) in order for the two techniques to be effective. Source inspection also will be presented in this chapter.

POKA-YOKES

The elimination of defects using *poka-yokes* is part of the just-in-time (JIT) philosophy and is included as one of the 20 keys (key number 11) of lean manufacturing, as shown in Fig. 4.1. Poka-Yokes improve the quality (reduce the defect rate) that is part of the overall equipment efficiency rate (Fig. 4.2).

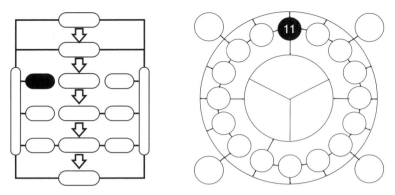

Figure 4.1. Location of *poka-yoke* in just-in-time and 20 keys diagrams.

THEORETICAL BASIS

Inspection and Statistical Quality Control (SQC)

Every production process generates defective products. The objective for all quality control systems is to reduce the number of defective parts. There are two types of defects: *isolated* (a crack appeared on a product) or *sequence* (repetitive) defects.

Product inspection is performed in order to avoid defective products from reaching the customer. These inspections can be classified according to different taxonomy:

- Total (100 percent of the products) or partial (extrapolating a sample study results)
- Statistical (based on statistical theory) or not statistical
- Quantitative (number of elements) or qualitative (appearance of the product)
- Based on measures (numerical values) or based on functional trials (without measures)

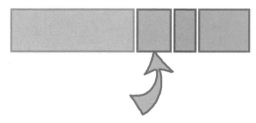

Figure 4.2. *Poka-yoke* utilization improves the quality rate.

- Sensory (carried out through human senses) or physical (by means of the use of devices such as gauges or meters)
- Subjective (carried out by the person that executes the production process) or objective (by another person)
- Internal (carried out inside the process) or external (like an independent process)

In each of the cases just described, the inspection should be able to discover defects and produce a corrective action in order to eliminate the errors that cause the defects; otherwise, inspection operations will become a pointless exercise.

Many years ago, it was believed that the only way to ensure the quality of all manufactured products was to inspect all the parts—100 percent inspection (Fig. 4.3). However, this procedure requires a lot of time and is not reliable. In an experiment, 100 faulty (defective) pieces were added to a "perfect lot." In the first inspection, only 68 defective parts were found. After 3 inspections, another 30 defective parts were found, and the last 2 defective parts were never found; proving that 100 percent inspection is not as effective as we think.

To avoid this problem and reduce inspection cost, it is possible, for example, to duplicate inspections (Fig. 4.4) so that mistakes made by the first inspector can be found by the second inspector. Unfortunately, defects were still showing up after duplication. Thus, how many control stages would be necessary to guarantee the quality of the entire manufactured lot?

It is evident that through this procedure it is not possible to avoid product defects in an efficient way. For this reason, statistical quality control (SQC) techniques seem to be the right alternative. This technique evaluates a product sample, and based on the number of defects found, a statistical decision to accept or reject the lot can be made. Unfortunately, some percentage of defects is always accepted, but the quantity of defects can be very small (nowadays six parts per million or less).

Figure 4.3. Can 100 percent parts inspection ensure the quality of the entire lot?

Figure 4.4. Can two inspectors ensure the quality of whole products?

In today's production, a high defect rate (high defect level) can be deadly for a company's well-being, and therefore, it is very important to eliminate or capture all defective parts before they reach the market. If in one million products the only defective part produced manages to reach the final customer, the company defect level from that customer's point of view will be of 100 percent.

It is important to mention that Shingo, in his book, uses the term *statistical quality control* (SQC), a term that today has become obsolete because it still refers to the quality control process. SQC has been replaced by *statistical process control* (SPQ).

From SQC to Zero Defects

Shingo introduced new ways of carrying out inspection processes that were based on SQC. The inspection process evolution that takes place at all firms should be similar to the historical evolution of these methods. SQC techniques are based on two principles: evaluate (statistical samplings) and inform (feedback analysis until reaching the cause/ process that created the defect).

SQC implementation in companies has two main problems. On the one hand, these methods do not guarantee the quality of all products (because 100 percent of them are not inspected), and the feedback and corrective actions to avoid such problems are slow or in many cases do not even exist. Therefore, an inspection process, which is based on control charts, does not reduce the factory defective rate; it only detects defects.

Shingo proposed two methods to avoid these problems:

- 100 percent inspection of the product using *poka-yoke* devices
- Accelerating feedback by self-checking, where production workers check their parts, and by successive check systems

Poka-Yoke. The *poka-yoke* concept was created by Shingo. At first, he called them "foolproof devices," but later he changed it to *poka-yoke,* which is Japanese for "unnoticed mistake proving." To consider an inspection device as a *poka-yoke,* it must be ingenious, simple, and cheap.

Several *poka-yoke* devices can be found in our activities of daily life. These devices are used to prevent errors from happening, e.g., the recording protection window on a floppy disk or the diameter of a gas nozzle that indicates to people the right gas to pump in order to avoid filling up a car with incorrect fuel (Fig. 4.5).

Several *poka-yoke* systems can be used in a company in order to mistake-proof activities. Some of the most common *poka-yokes* used are automatic part feeders, which guarantee that 100 percent of the pieces are separated correctly. In addition, *poka-yokes* can reduce undesired workload components. If a worker is in charge of performing the mentioned task, it would be very exhausting for him or her to feed components constantly.

Figure 4.6 shows a typical *poka-yoke* example where each part (depending of its height or width) is separated into different boxes. Through *poka-yoke* devices, it is possible to separate parts with different specifications (different heights, different box). The higher parts are placed into the first box, and the shorter ones continue to the end of the conveyor belt. *Poka-yoke* devices such as "go, no go" devices are used to avoid inspections based on trials.

Self-Check and Successive-Check Systems. It is not always possible to design a *poka-yoke* to carry out 100 percent of the inspections, and therefore, it is necessary to inspect parts conventionally.

Figure 4.5. Different diameters in gas hoses can be considered a *poka-yoke* device.

Figure 4.6. Poka-yoke example.

In a *self-check procedure,* the worker who produces the part is the same worker who carries out the inspection. This system is the most efficient one because the worker obtains immediate feedback concerning the manufacturing process. However, it may be necessary to provide additional inspections because criticizing one's own performance may not be totally objective.

In the *successive-check procedure,* the next worker (the worker who comes right after the worker who produced the part) typically will conduct the inspection task. The successive-check procedure, according to Shingo, can reduce the defect rate by as much as one-fifth the initial value in about one month. To achieve this result, it is necessary to fix two or three check points. It is also important to keep in mind that in the beginning of this checking procedure, the defect rate will increase because many of the process defects that went undiscovered previously are going to be detected. In the case of sensory-based inspections (i.e., scratches, painting quality, etc.), it is advantageous to place samples (acceptable and not acceptable) next to the checking point to show acceptable limits.

Source Inspection. The techniques just discussed reduce the defect rate but do not eliminate the errors that produce the defects. There is a causal relationship between errors and defects (Fig. 4.7). Therefore, if the error source is eliminated, errors will never become defects, and a defect-free process will be achieved.

Through source inspection, all the errors are eliminated except the unnoticed errors or unavoidable plastic swarf from casting or injection molding (e.g., part fins or leftovers). These unnoticed errors can be detected by an efficient use of *poka-yoke* devices.

Figure 4.7. Source inspection eliminates the errors that produce the defects.

There are two types of source inspection:

* Vertical: before the process
* Horizontal: inside the same process

POKA-YOKE DESIGN METHODOLOGY

To design a *poka-yoke,* the three methods illustrated in Fig. 4.8 can be employed. *Poka-yoke* design can be based on the weight, the dimensions, or the shape of the element.

Other strategies also can be used, such as meters/counters, the spare pieces method, or a fixed sequence in the assembly process (Fig. 4.9).

Once a certain method or procedure for detecting defects is determined, it may be advantageous to use technology to design the *poka-yoke* devices. Contact mechanisms (limit switches), mechanisms without contact (sensors), and meters/counters are examples of suitable technology (Fig. 4.10).

Poka-Yoke Examples

This section provides some examples of *poka-yokes.* The common characteristic of these devices is that all of them are simple devices.

The problem with the process shown on the left side of the Fig. 4.11 is that improper parts (extra material) can break or otherwise adversely

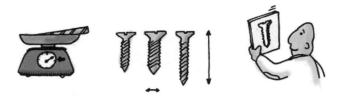

Figure 4.8. Three suggestions to design a *poka-yoke:* limit size, weight, or volume.

Figure 4.9. Spare pieces indicate that something was forgotten during product assembly.

affect the tool of the following process. In a case such as this, it might be necessary to inspect or gauge all the manufactured products on the fly (not ideal).

The introduction of a mechanical blade stop as shown in the figure can redirect the parts or even stop the line when something goes wrong and sound and alarm for immediate intervention. It is possible to prevent the line from stopping by using a "size limit" *poka-yoke* (the blade deflector) to make sure that parts of improper height or diameter are redirected, as shown on the right side of the Fig. 4.11.

Figure 4.12 shows how certain templates (designed previously) eliminate errors when adjusting the parameters before a process. By designing separate face covers for the proper product, the template patterns indicate the proper settings and values that should be used.

Figure 4.13 shows how a simple slot in a conveyor line can be used to remove an incorrect product (incorrect position, upside-down part) from continuing through the production line. Without the *poka-yoke,* filling material can be wasted, and a major spillage can occur. By using *poka-yokes,* the line will not stop, and the improperly oriented parts will fall to a collector box (because they do not have the proper ori-

Figure 4.10. The use of sensors allows one to design a *poka-yoke.*

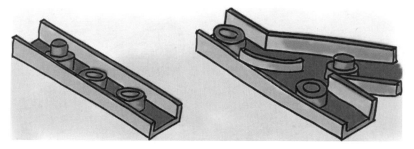

Figure 4.11. Example of a *poka-yoke.*

entation geometry or potentially enough weight to stay in the production line).

SUMMARY

This chapter had the primary objective of presenting one of the many quality-related improvement tools: the *poka-yoke* (unnoticed mistake-proving device). The lean manufacturing philosophy considers this tool as one of the pillars of improving overall equipment efficiency. This chapter has presented several examples of this kind of device, showing that in order for a device to be considered as a *poka-yoke,* it must be ingenious, simple, and cheap.

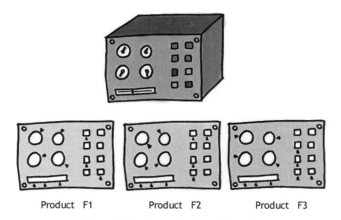

Product F1   Product F2   Product F3

Figure 4.12. Example of a *poka-yoke.*

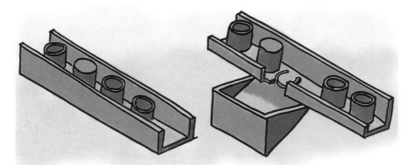

Figure 4.13. Example of a *poka-yoke*.

RECOMMENDED READINGS

Nikkan Kogyo, *Poka-Yoke: Improving Product Quality by Preventing Defects.* Pittsburgh, PA: Productivity Press, 1988.

Shigeo Shingo, *Zero Quality Control: Source Inspection and the Poka-Yoke System.* Pittsburgh, PA: Productivity Press, 1986.

5

Equipment Efficiency: Performance and Motion Study

In this chapter we will examine the second equipment efficiency indicator based on two factors: equipment performance losses owing to minor stoppages (usually not registered) and equipment performance reduction caused by equipment component deterioration and/or wear. In Chap. 7 we will study equipment availability (setup reduction) and quality related to startup.

Small breakdowns or device (fixture and tooling) holdups are responsible for these machine stoppages. In other cases, an improper adjustment or interaction between the worker and the machine cycle also can create problems.

Time and motion study allows one to optimize the relationship between the worker and the machine, as well as to investigate whether the worker can tend more than one machine in those cases where the machine cycle is significantly longer than the worker cycle.

The main tools presented in this chapter are worker-machine and machine-machine diagrams. These tools help us to study the relationship between the worker and machine cycles (or between machines), eliminating or reducing idle time and optimizing the working cycle.

MOTION STUDY

Motion study (also known as *movements study*) is defined as the methodology whose goal is to decrease the amount of work (shorter working cycle) by improving the existing work methods and layout. The result of a well-conducted motion study is that idle times are decreased or eliminated completely, i.e., those nonproductive activity times where non-value-added activities (walking, searching for, etc.) are eliminated from the process.

If this methodology is applied to the machine setup process, it could be considered a single-minute exchange of dies (SMED) tool, as will be shown in Chap. 7. Motion study is included in the 20 keys of lean manufacturing as key number 6 (Fig. 5.1). In addition, this tool improves the overall equipment efficiency ratio (Fig. 5.2).

Time study, as explained in Chap. 3, is one of the first steps in a motion study because in most of the cases it will be necessary to quantify the current standard as well as to document any time savings. In addition, it is necessary to have available some approximate times in order to carry out the analysis.

In a motion study, alternate approaches (methods) to performing a production operation are proposed and then analyzed. The new procedure at first may seem worse than the current method, but this can be due to the lack of familiarity and the change resistance of workers who are accustomed to a personal way of carrying out tasks. Shingo illustrates resistance to change with an explicit example of animal behavior. Shingo shows, as illustrated in Fig. 5.3, that a fish will not change its learned behavior of selecting a path through a previous obstacle to obtain food even if the obstacle is removed. The fish follows the same path instead of reducing the distance to its goal by going straight ahead. Although this is a simple example, change is usually met with resistance in much of our everyday activities.

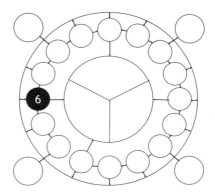

Figure 5.1. Location of motion study in a 20-keys diagram.

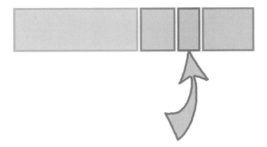

Figure 5.2. Motion study improves the performance rate.

Motion study also can be used to design a new work procedure (instead of improving an existing one) or to eliminate variability by changing the way a task is conducted. The importance of eliminating minor stoppages depends on the process, its complexity, its relationship with other upstream or downstream processes, and the *takt* and cycle times. For this reason, a 3-second cycle reduction can have a major impact on a process and should not be summarily rejected without further investigation. For a *takt* and equivalent cycle time of 2 minutes, a 3-second reduction in a day of 24 working hours would result in the production of 18 more parts. This, of course, assumes that this station will still have the maximum cycle time and that all saving go directly into the making of new product.

For a more complete treatment of the reduction and complete elimination of minor production stoppages, the reader is referred to the Recommended Readings at the end of this chapter. For example, Kikuo Suehiro discusses methodologies for eliminating minor stoppages. These methodologies will not be presented further in this book.

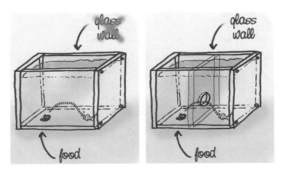

Figure 5.3. Habit change is the biggest obstacle to new method implementation.

THEORETICAL BASIS

Motion Economy Principles

In Chap. 3 we discussed the work of Frank and Gillian Gilbreth and how they defined the operation standard time by listing and measuring working elements. When analyzing a complete operation, they divided it in small tasks called *therbligs* (their last name in reverse). They regrouped and optimized these small tasks to form the complete operation in various ways.

The principles of motion economy are the result of their diligent work, and these principles are still used. Thanks to their work, Frank was able to outperform younger bricklayers theoretically more qualified than him. Figures 5.4, 5.5, and 5.6 show that these principles of motion economy are simple rules that facilitate the proper design of operations. In each figure, the drawing of the left represents a nonefficient process.

For example, if a boiler is to be controlled manually so that it does not exceed a certain pressure, it is recommended to use analog displays with properly noted performance zones rather than digital ones (Fig. 5.7). Other examples make reference to the efficient use of other displays. It is recommended, for example, lateral movements be used instead of up-down movements.

It is important to distinguish the principles of motion economy (whose main objective is to increase productivity by means of improving the method of carrying out tasks) from other studies that analyze the best working conditions to increase productivity. For example, some studies have demonstrated that productivity increases when work is done under poor lighting conditions. It is not necessary to comment on the negative consequences that this situation would have on the worker motivation. Since Gilberth's studies, the human body has been analyzed in detail, and optimal working distances and work envelopes have been

Figure 5.4. Better to use symmetrical movements.

Figure 5.5. Better to split the weight.

defined. For example, arms-sweeping areas have been fixed when the worker is seated or standing (Fig. 5.8).

Proper bookcase heights for storage are also known (Fig. 5.9). In this way, accidents can be avoided or reduced. In some cases, the hands can be let free by using feet to perform certain tasks (e.g., a foot activation pedal to close a vise). Predetermined time, which was studied in Chap. 3, was derived from these type of studies.

MOTION STUDY TOOLS

Value Analysis

This analysis focuses on worker movements and workplace layout. Applying the motion economy principles described earlier, several production tasks normally can be redesigned and improved. Some companies have an ergonomics department, where the operations performed in the plant are analyzed and improved routinely (Fig. 5.10).

As is shown in the following figures, important improvements in workplace layout and methods can be achieved, and unnecessary move-

Figure 5.6. Better sitting down.

Figure 5.7. Better to use an analog control.

ments can be eliminated. As shown in Fig. 5.11, the only difference in the two activities is the location of the stop valve. The time required to activate the valve, as well as worker safety associated with closing or opening the valve, can be improved with a minimal change in the stop valve location. By analyzing body movements and improving equipment design, worker efficiency can be improved significantly.

Figure 5.12 shows how to reduce the effort required by a worker if an electric or pneumatic hoist is installed. Even if the weight is not excessive, the number of times that this task is carried out during the day makes this process very harmful for the worker.

5W2H and 5-Why Methods

The 5W2H tool is very simple and consist on asking a group of questions about the task that is being analyzed. The tools and process are illustrated in the following and in Fig. 5.13.

Frequently, the response to change is about the same. Workers seldom see the need to think about another method for carrying out a

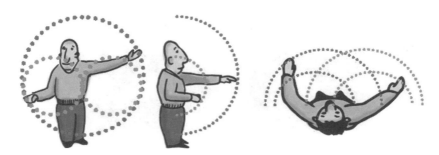

Figure 5.8. Working distances for the arms.

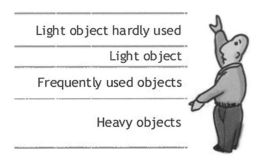

Figure 5.9. Ideal classification for bookcase height.

task. This technique facilitates the analysis of operations that do not add value to the process.

Although some of the solutions seem trivial, it is routine to find situations such as the following examples that are far from reality. Figure 5.14 shows the old and the new methods to unload a truck.

Is it also useful to employ another method, called the *5-why method.* This method consists of asking "Why?" up to 5 times in order to find the root of the problem. Asking yourself "Why?" 5 times helps to differentiate the symptoms that produce the problem from the real causes of the problem (the ones that needs to be eliminated).

Worker-Machine Diagram

In the situations described earlier in the value-analysis section, direct observation or the use of a camcorder is enough to document a problem for analysis. Sometimes, however, it is necessary to analyze a very fast or a very slow operation. In the first case, there are camcorders that are able to carry out up to 960 shots per second. As a result, unnec-

Figure 5.10. Value analysis improves factory operations.

Figure 5.11. Reinstalling the stop valve. Which one looks best?

essary movements can be detected. For the second case, if the operation that needs to be studied takes up too much time, then slow camcorders (one shot per minute) also can be used, e.g., the ripening process of an apple.

These two particular cases are rarely encountered in the industrial environment. It is more common to use production cycle analysis in which the machine and the worker interact with each other. In order to carry out a proper study, the operation should be divided into short but measurable tasks, separating machine and worker tasks.

The use of the worker-machine diagrams can facilitate this work significantly. These diagrams represent the task sequence that the worker carries out when interacting with the machine. Such diagrams also are used to analyze idle time and are considered an important improvement tool (Figs. 5.15 and 5.16).

In analyzing the worker idle time using the worker-machine diagram after the data are collected, the analyst is able to optimize the working

Figure 5.12. Pneumatic elevator reduces the needed effort.

5W and 2H	Questions
What	What is this operation for?
	What would happen if it is not performed?
Who	Who does the operation?
When	When is it done?
Where	Where is it done?
Why	Why is it necessary to do it?
How	How is it done?
How much	How much does it cost to do it?

Figure 5.13. The 5W2H tool.

cycle. By rearranging or modifying the worker-machine diagram, cycle time can be improved. As a result, the machine idle time decreases, as is shown in the new worker-machine diagram in Fig. 5.16.

Machine-Worker Ratio

If the task duration of the machine is very long, the worker could operate more than one machine. The right number of machines controlled by each worker is determined through the *machine-worker ratio:*

$$\text{Machine-worker ratio} = \frac{\text{machine cycle time}}{\text{worker cycle time}}$$

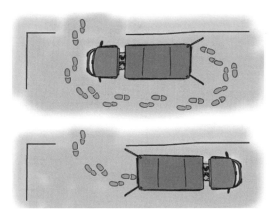

Figure 5.14. The answer to the "Why" question was, "We always do it like this."

Worker	Machine
Insert part	occupied
	Processing
Remove part	occupied
Inspect part	
Introducir pieza	occupied

Figure 5.15. Worker-machine diagram of the initial situation.

In the machine cycle time, all those operations where the machine is operating or occupied are considered, even those shared with the worker. The same consideration must be taken into account when analyzing the worker cycle time. This includes the worker tasks and those tasks that affect the worker and the machine simultaneously.

The value of the ratio determines two possible situations for the right number of machines that a worker can handle simultaneously. For example, if the ratio is 7.6, the worker should be able to handle 7 or potentially even 8 machines. Determination of the manufacturing costs in each case will fix the number of machines the worker should handle.

Worker	Machine
Insert part	occupied
Inspect part	Processing
Remove part	occupied
Insert part	occupied

Figure 5.16. Worker-machine diagram of the final situation.

For example, Fig. 5.17 shows a worker-machine diagram when a worker is handling two or three machines. The figure represents the working cycle from the start of the process until it reaches a steady state. The machine-worker ratio must be studied in this steady state.

If one observes transient activities (or temporary evolution until a steady state of operations that the worker and the machine carry out is achieved), the following important conclusions can be made:

- In the case where the worker handles two machines (diagram on the left in the figure), the working state is fixed by the machine, since the time that the worker needs to carry out all the operations on the two parts produced is shorter than the machine cycle. Therefore, the worker will wait 1 minute until the first machine finishes before he or she is able to load the next part.

- When the worker handles three machines (diagram on the right), the necessary time to carry out the operations on the three parts is longer than the machine cycle, and therefore, each machine will finish processing the part before the worker completes his or her activities. As a consequence, each machine will have to wait one minute until the worker finishes the work on the third part.

Machine-Machine Diagram

One-piece flow strategy presented in Chap. 2 pointed out the importance of achieving an ideal flow between working centers. In this case, the relationship between machine operations involved in the manufacturing process should be studied in depth.

The machine-machine diagram, similar to the worker-machine diagram, can be used to analyze product flow and the idle time in one working cycle when two or more machines are related.

In order to draw the diagram (always in a steady state), it is important to start with the machine that fixes the working cycle, because this machine does not have idle time. There is no methodology to complete the rest of the diagram, but the rest of the machines' operations can added in an attempt to minimize product lead time.

SUMMARY

In this chapter, some tools to facilitate motion study in a company have been presented. Worker-machine and machine-machine diagrams facil-

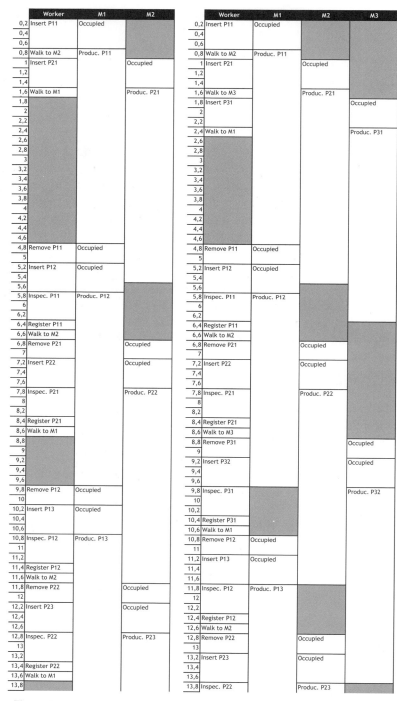

Figure 5.17. Machine-worker diagram from the beginning to the steady state.

itate identification of idle time in working cycles. Thanks to the improvement methodologies presented, it is possible to eliminate those idle times. Therefore, the performance rate (included in the overall equipment efficiency ratio) increases.

RECOMMENDED READINGS

Kikuo Suehiro, *Eliminating Minor Stoppages on Automated Lines.* Pittsburgh, PA: Productivity Press, 1992.

Tomo Sugiyama, *The Improvement Book: Creating the Problem-Free Workplace.* Pittsburgh, PA: Productivity Press, 1989.

6

Equipment Efficiency: Availability, Performance, and Maintenance

The role of maintenance is to ensure the survivability and proper functioning of all company hardware (productive and nonproductive). Most maintenance departments are considered, by most companies, "a necessary evil" or a money pit that represents a continuous cost. Managing a maintenance department at times can be nearly impossible because the investments required to improve production processes usually take on a low priority or, even worse, may not even make it to the priority list for capital expenditures.

Maintenance evolution, as well as maintenance techniques evolution, has been developed in parallel for many companies: The first obligation of the maintenance department is to remediate hardware failures that have occurred already. The next obligation after fixing breakdowns is to prevent future problems with the equipment that eventually may lead to failure.

The companies that are most advanced in maintenance management try to incorporate basic maintenance tasks into their daily production routine such that direct labor personnel check fluid levels and examine production equipment for potential failure mechanisms, also searching for ways to increase the ability to predict potential equipment breakdowns.

EQUIPMENT MAINTENANCE

The primary objective for the maintenance department of any company is to maintain the productive resources at a high operative level in order to ensure their service at an expected cost. There are several terms that attempt to explain why equipment maintenance is important. These terms are helpful, but no expression is as clear as the one that defines maintenance as the "machine's medicine" (Fig. 6.1).

According to this analogy, the maintenance department is in charge not only of fixing the machines (or correcting improper performance) but also of carrying out checkups concerning the machines' well-being in order to avoid breakdowns (prevent improper performance). The maintenance mission is to keep the equipment in good working order and also to determine the right moment to replace equipment. When the predicted replacement time is reached, the effort to keep the machine running is no longer cost-effective, and it is better simply to take it offline.

The elements that need maintenance in a factory are numerous. They include direct (machine tools, welders, etc.) and indirect production equipment (test equipment, coolant distribution equipment, etc.). Basically, all available equipment requires maintenance, even if it is not used directly to carry out productive tasks:

- Machines (mechanical, electrical, and pneumatic parts) and tools
- Facilities (compressed air, heating, electrical systems, etc.) and buildings (walls, illumination, etc.)
- Information and transportation systems (if they belong to the company)

Some companies subcontract maintenance, whereas other companies manage maintenance of their own resources. A good example of this situation is the fact that most companies now subcontract their facilities

Figure 6.1. Maintenance is the "machine's medicine."

and building maintenance, e.g., mowing the lawns, general landscaping, and parking renovations. In cases, the maintenance service is fully sub-contracted. In other cases, there is no on-board staff to deal with machine breakdowns.

Efficient maintenance management and equipment conservation are contemplated in both just-in-time (JIT) and the 20 keys (key number 9) for lean manufacturing (Fig. 6.2). Breakdown reductions improve the availability and performance rates of the equipment (Fig. 6.3), thereby improving overall equipment efficiency.

THEORETICAL BASIS

Types of Maintenance

All industrial equipment is exposed to transitory (wear) or definitive breakdowns (catastrophic failure) affecting its functionality and performance. In many cases, equipment failures can represent high costs for enterprises, as well as high risks for the workforce; therefore, they must be solved or repaired as soon as possible.

However, the maintenance mission is not just to repair the breakdowns. Maintenance should be able to get ahead of the breakdowns. Both tasks correspond to the two major types of maintenance: corrective and preventive.

Preventive maintenance has two variants:

- Systematic preventive maintenance
- Conditional preventive maintenance or predictive maintenance

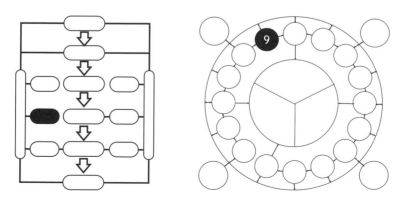

Figure 6.2. Location of maintenance in just-in-time and 20 keys diagrams.

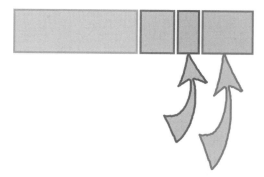

Figure 6.3. Maintenance improves availability and performance rates.

Corrective Maintenance. Until the 1950s, corrective maintenance was virtually the only maintenance carried out in companies, also called *breakdown maintenance*. During those years, machine stoppages hardly affected productive time. The machines were less complex and more reliable, and the repairs, mainly in the mechanical parts, were carried out in an effective way.

Figure 6.4 illustrates the time evolution for the level of performance of a particular piece of equipment when applying corrective mainte-nance only. There are two types of corrective maintenance:

- *Urgent repairs.* This type of repair is for machines that have had a breakdown, causing the machine stop. In most cases, returning

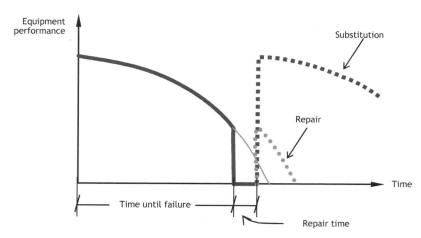

Figure 6.4. Behavior of a component with corrective maintenance.

the equipment to service has a higher priority than fixing the machine properly. In other words, the repair that is carried out is provisional one, and the machine downtime should be as small as possible. The remaining tasks needed to repair the machine fully will be scheduled for a future time.

- *Scheduled corrective.* This type of maintenance appears as a result of urgent repairs. When machine service is reestablished, it is necessary to determine an appropriate time to repair the machine completely. After repairing the damaged component, it can be as good as new or at least as it was before (e.g., replacing a light bulb or repairing a flat tire).

The main problems with performing repair tasks are

- Repair tasks are performed quickly and under pressure, which can cause future problems.
- Repair time can be very high because replacement part(s) needed to get the machine running again may have to be ordered from a supplier.
- Accidents can take place because of poor maintenance safety measures.

Keeping only a corrective maintenance policy implies higher labor costs, especially when several machines are damaged at the same time. In this case, enterprises have to deal with high-idle-time issues. However, this policy can be justified in some cases, e.g., when maintenance is only performed on equipment with a frequent replacement policy (such as office computers) or when breakdown costs are small (light bulbs fail).

Systematic Preventive Maintenance. In the early 1960s in the United States, General Electric Corporation systematized a new type of maintenance called *planned maintenance*. This type of maintenance focused on improving certain machine aspects after breakdowns had been repaired. Four years later, planned maintenance arrived in Japan, where the bases of this systematic preventive maintenance process were established. At first, this maintenance consisted of only a systematic substitution of some machine components (that still were working properly) with new components.

This type of maintenance is applied to general wear or use components, such as bearings and filters, and is used on equipment with high failure costs. This systematic replacement policy is carried out to avoid breakdowns. Besides the systematic replacement, it is also very important to know with precision the component's performance characteristics (curve) in order for this type of maintenance to be effective (Fig. 6.5).

Component replacement can be carried out continuously. For example, a typical replacement policy might call for a component to be replaced every week (even though this method is getting obsolete) or in other ways, such as every 300 working hours or every 1000 parts produced.

Conditional Preventive Maintenance or Predictive Maintenance. Systematic preventive maintenance can become very expensive because several components can be replaced despite still being in good operationing condition. Conditional preventive maintenance is a method that is used to change components based on their current state (Fig. 6.6). Using this maintenance policy, the useful life for costly components can be extended.

For conditional or predictive maintenance, it is not critical to have very accurate component performance curves; however, this type of maintenance best fits components for which performance can be monitored (via product attributes) or judged by the operator. It is therefore necessary to have a strict and detailed method for analyzing process data, noticing the actions required for specific conditions. ISO 14,000 (corresponding to the environmental norm) requires that industry avoid systematic preventive maintenance when working with oils and envi-

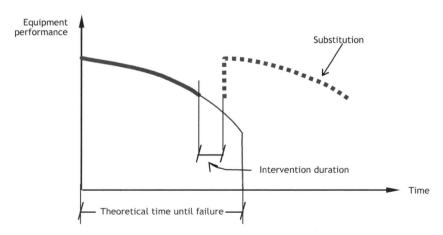

Figure 6.5. Behavior of a component with systematic preventive maintenance.

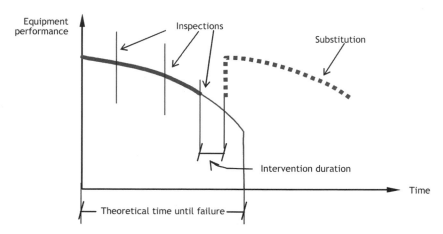

Figure 6.6. Behavior of a component with conditional preventive maintenance.

ronmentally harmful products and encourages conditional or predictive maintenance.

Sometimes, it is possible to monitor, in real time, a component's state and determine when working anomalies are about to occur. In this case, maintenance workers must proceed with immediate component replacement. Moreover, such maintenance generally is called *predictive maintenance.*

The principal difference between conditional preventive maintenance and predictive maintenance is that the operating variables are adjusted constantly using conditional predictive maintenance (large families of parts may be produced). It becomes necessary to look for correlations between multiple parameters and the degradation of a component.

Some variables that can be measured are temperature (thermocouples), noise (phonometer), cracks (x-ray machine), and pressure losses (manometer). The normal QS9000 recommends predictive maintenance procedures, and in the case of not having them, it is necessary to justify the economic problems of investing in them.

MAINTENANCE PROGRAM IMPLEMENTATION

Prior to selecting the best methodology to improve equipment availability (reducing the number of breakdowns), it is always a good policy to first find the root of the breakdowns, as well as the current state of the company (maintenance-wise). Almost all machines follow a similar life cycle from installation until disposal:

- *Hidden small defects.* It is difficult to detect the appearance of these defects because they are hard to see and are not apparent at first. This type of defect normally is unattended and most likely will go unfixed owing to the fact that it does not interfere with functionality. An example would be increased friction in an axle.
- *Apparent small defects.* These types of defects are more noticeable, e.g., small vibrations on a machine. When the defects do not affect the equipment's functionality directly, they are normally not repaired either.
- *Execution under expectations.* In this case, the defects affect equipment productivity. Most likely the machine will not work accurately, and the standards of quality will be violated. Often it is difficult to find the problem or cause because the specific damaged component is unknown.
- *Intermittent stops.* The machine intermittently produces defective parts, and there are numerous process settings. Small repairs that get the machine working are performed, but in a provisional manner.
- *Stops and breakdowns.* The machine's performance is poor, and breakdowns are frequent. Breakdowns can become expensive in terms of lost production time and money (Fig. 6.7).

In some companies, production equipment may not be as new as one would like it to be. This equipment may have already gone through some of the stages just described and most likely is at the "stops and breakdowns" stage. Replacement or renewal (rebuilding) should be considered here.

Another problem that firms face is that production equipment typically becomes more sophisticated, faster, and more expensive each year (i.e., longer payback or extended capitalization). The result of pur-

Figure 6.7. Poorly maintained equipment.

chasing new equipment is that the equipment has greater economic impact (higher product volume), and therefore, repairs should be done at a faster rate. Working shifts also can represent an obstacle to maintenance interventions. The implementation of working shifts (up to five in some cases) limits possible maintenance tasks and scheduling.

The objective of maintenance is to efficiently oversee equipment throughout the equipment life cycle through all the stages described in this section. In order to be successful, it is necessary to cover the entire life cycle, beginning with implementing an effective corrective maintenance plan, then preventive maintenance tasks, and later on, predictive maintenance strategies.

Getting Started

The first step, before any maintenance policy implementation, is to become familiar with the resources that will require maintenance. Each maintained resource should be coded (resource ID) in order to distinguish it from the rest of the resources because it is not unusual to have identical pieces of equipment.

It is also important to code the types of breakdowns, as well as the maintenance tasks performed. As a result, in a historical data study, it is possible to group failure causes of each piece of equipment in order to take proper corrective actions. There is a specialized taxonomy that helps to code resources and breakdowns, although most of the time common sense and simplicity lead to a good coding system.

After this preliminary phase, maintenance should have the following two documents:

- *Facility inventory.* This document lists all the plant equipment and its principal characteristics: code, record number, equipment type, etc.
- *Equipment history files.* This is a paper (or electronic data) portfolio that contains the data given by the equipment manufacturer, information about location of the equipment in the plant, an outline or a picture of the equipment, the characteristics of the equipment installation, and finally, the types of spare parts needed.

Corrective Maintenance Implementation

For corrective maintenance, the number one priority is to organize, in an effective way, the corrective maintenance procedures and actions.

When a breakdown occurs, the worker who has discovered it should fill out a breakdown work order. This document will be different in each facility because the information needed in each maintenance department varies.

If the worker can solve the problem, he or she must fill out a report to the maintenance department, so that the breakdown gets registered; otherwise, the work order will be sent to the maintenance department, and a work request order will be issued.

Depending on the work urgency, maintenance workers either will repair the machine immediately or will schedule the repair. The repair can be provisional or definitive. In the first case, it is recommended to keep the work order open until the repair is completed.

Scheduled Corrective Maintenance. Variability in the duration of corrective maintenance tasks can be problematic if they are not scheduled properly. In order to complete repairs efficiently, corrective orders and flow diagrams for repetitive repairs must be developed explaining how the repair should be done and listing the materials and spare parts that should be used. Adequate maintenance time should be allocated for all repairs.

When maintenance workers carry out corrective operations, the task does not end with the equipment repair. There are some other procedures that must be followed. Maintenance workers should gather all the breakdown information (Fig. 6.8) and describe the process that was performed to solve the problem and document them in the work order.

Each machine should have its own file with breakdown records; otherwise, it can be very difficult to analyze breakdown causes and anticipate future problems accurately. This file must be upgraded with each maintenance intervention; therefore, the work of maintenance workers does not end when the resource is running again but when they have fully registered the breakdown in the records. For frequent airline trav-

Figure 6.8. Maintenance workers should gather all the breakdown information.

elers, this action is the all too familiar process of fixing the switch or the signal light in the cabin (5 minutes) and then waiting while all procedures and issues are signed off and cleared (20 minutes). In cases where life is at risk (as in air travel), repair, report, acknowledge, and then recheck constitute common practice. Corrective maintenance tasks consist not only of changing the broken or malfunctioning components but also of studying the causes and frequency of the breakdowns. This is the most appropriate time to introduce into a company the concept of preventive maintenance.

Preventive Maintenance Implementation

Of all of the numerous goals of preventive maintenance, the number one concern is to avoid a breakdown of any resource while keeping maintenance cost as low as possible. Preventive maintenance includes two types of actions:

- *Inspections.* The objective of this action is to observe and detect possible anomalies. Generally, inspections consist of frequent checkups, sometimes daily, that follow a specific inspection plan.
- *Revisions.* These actions require scheduled equipment stops and usually suppose a systematic substitution of several machine components.

In companies with machines that run in sequence, scheduled stops usually are carried out over the weekend (or during nonproductive periods), increasing maintenance costs. Preventive maintenance task scheduling is mandatory in the ISO norms. These tasks can be planned daily, weekly, monthly, or even annually. They always should be scheduled at times when they do not affect the factory's production plan. However, the reality is that daily working problems force us to continuously reschedule these tasks. It is still worthwhile to schedule these preventive maintenance tasks and to try to adhere to this schedule.

Equipment preventive maintenance tasks are also called *PM orders.* Each PM order should be based on a study of the causes of equipment breakdown. This study can be carried out by using the equipment FMEA tool described in the tools section later in this chapter.

Autonomous Maintenance

Maintenance staff normally will carry out most of the PM orders; however, some PM orders can be carried out by production workers as part of batch setup or shift startup. This group of orders is known as *user maintenance orders.*

These user maintenance orders are the key in the development of autonomous maintenance, and they should be simple and graphically represented, indicating to the workers the exact place of all the elements that should be inspected and the tasks that should be carried out (Fig. 6.9).

Again, the calendar for autonomous maintenance repair tasks (self-employed maintenance) is basic because many inspection tasks should be carried out every day. Therefore, the maintenance department may receive a significant number of notices that could be handled easily by production worker (e.g., a dirty optical switch, a plug that does not work, etc.). In some cases, it takes more time to fill out the request order than to fix the problem. Autonomous maintenance includes these small tasks and three daily preventive measures: cleaning, lubricating, and checking (Fig. 6.10).

Security for autonomous task execution is one of the most important restrictions when determining the type of activities that production workers can perform safely (Fig. 6.11). Repair or maintenance never should be performed by the operator if the knowledge required to fix the machine is high. For complex maintenance, a repair specialist should take care of problem. Autonomous tasks apply only for simple repair operations. The maintenance department staff should perform complex maintenance tasks as well as dangerous tasks.

Sometimes it can be very challenging to convince production workers about the importance of maintenance tasks. Companies face this issue because workers do not want to perform tasks that they do not consider their responsibility. This situation is similar to checking the oil in a car. Most of the time, the oil level is okay, and the check is

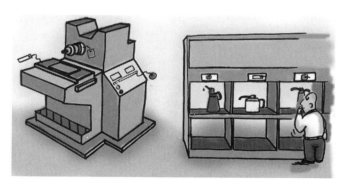

Figure 6.9. User maintenance orders must be easy for production workers to understand and accomplish.

Figure 6.10. Cleaning, lubricating, and checking the equipment.

not necessary. Even worse, the maintenance department may not allow operators to perform maintenance tasks on the equipment.

The autonomous maintenance process has a specific methodology (Fig. 6.12) made up of seven steps that lead to its complete implementation.

TPM: Total Productive Maintenance

In the 1970s, Nakajima developed *total productive maintenance* (TPM) in Japan. TPM was a new maintenance management philosophy included in Toyota's improvement process.

Although the TPM concept was introduced in Japan in 1971, it took a long time to be published in English. In fact, the first English translation was not published until 1988. TPM development resulted in the creation of the Japanese Institute of Plant Maintenance (JIPM), which grants a PM prize to top companies in TPM. Sixty percent of the winning companies over the first 17 years are now part of the Toyota group or suppliers of this group.

Nakajima combined preventive maintenance theories with the total quality concept. As a result, Nakajima developed the overall equipment

Figure 6.11. Security is one of the main restrictions in autonomous task assignment.

⑦ Autonomous Supervision

⑥ Process Quality Assurance

⑤ Autonomous Maintenance Standards

④ Overall inspections

③ Cleaning and Lubricating Standards

② Countermeasures to Sources of Contamination

① Initial cleaning

Figure 6.12. Autonomous maintenance implementation methodology.

efficiency (OEE) ratio (studied in Chap. 1), which is one of the TPM keys:

- *Maximize the overall equipment efficiency.* This is done by eliminating the six big losses described in Chap. 1 (breakdowns, setup and changeover, idling minor stoppages, reduced speed, defects and rework, and starting losses).
- *Autonomous maintenance implementation.* This is done in order to terminate the "I operate, you repair" mind-set.
- *Preventive engineering.* This is done to avoid the need to carry out maintenance operations on equipment, improving its maintainability (see reliability concepts in the tools section of this chapter).
- *Training workers for maintenance improvements.* Workers who operate the machines should be able to propose methods for increasing equipment availability by eliminating breakdowns or expediting repairs.
- *Initial equipment management.* This is done in order to avoid negative effects from the machine setup process.

Nakajima proposes specific objectives in each category of equipment losses and develops, mainly, the activities related directly to machine maintenance. All the PM prize-winning companies have an OEE ratio greater than 88 percent.

Although setup time and initial run quality, to mention but a few targets, affect a machine's efficiency, they are addressed in detail in another specific methodology that will be explained in a later chapter. In the maintenance field, the objective of TPM, or the ideal situation, is *zero breakdowns.* To achieve this goal, it is necessary to use such tools as P-M analysis, which is presented in the maintenance tools section later in this chapter.

RCM: Reliability-Centered Maintenance

RCM was created in the United States in the 1960s to optimize the reliability of aeronautical equipment. RCM was not used in nuclear power stations until the 1980s (after the Three Mile Island accident) and has been implemented only recently in the rest of the industrial world.

In order to begin RCM implementation, it is necessary to have a complete maintenance and breakdown record for each piece of equipment. Starting from this equipment information, the RCM objective is to determine maintenance tasks that are most effective for critical components. There is a specific methodology that facilitates RCM's implementation in companies based on several tools, such as FMEA, reliability analysis, statistical techniques, etc.

To apply RCM efficiently, it is necessary for a company to have a preventive maintenance program that runs properly. The starting point for RCM is performing a statistical analysis of equipment behavior owing to breakdowns; thus, without enough statistical information to characterize breakdowns, RCM cannot be applied adequately.

MAINTENANCE TOOLS

FMEA for Equipment

All defects have a root cause, and to eliminate future defects, an action must be carried out. For example, if the defect is in the gap between two elements (contact between them instead of clearance), then the cause or root of the problem can be lack of lubrication or a loose fastener. The action in each case would be different (i.e., grease or tighten the lose element).

To establish a good preventive maintenance plan, all the possible breakdowns, their causes, and their corrective actions must be analyzed. The main tool to carry out this type of analysis is failure mode and effects analysis (FMEA) for equipment.

The FMEA tool (Fig. 6.13) is a guide to analyzing, in an organized manner, causes of possible equipment breakdowns. In order to avoid breakdowns, a group of workers is gathered to study the problems and failures that can take place on the equipment. This team establishes action plans to avoid the failure causes discovered.

In an FMEA study, the team should enter data for each of the fields corresponding to each column in the template (see Fig. 6.13):

EQUIPMENT _____ CARRIED OUT BY _____ DATE _____ PAGE ____ OF ____

Equipment functions	Failure modes	Failure efects	Severity	Criticity	Failure causes	Probability	Actual controls	Detection	R.P.N.	Recommended actions	Responsible

Figure 6.13. FMEA template.

- *Equipment functions.* In this column, all the functions that the equipment carries out are registered. For example, (1) manufacture specific parts, (2) provides compressed air during specific conditions, etc.
- *Failure modes.* Through brainstorming, all the possible ways that the equipment can be forced to stop (breakdowns) are determined. Each stop is related to one of the functions previously registered, e.g., breaks, blockage, leaks, etc.
- *Failure effects.* All possible consequences of each failure are analyzed in detail. The most important factors to analyze are the effects and failure severity S for each failure. The impact or consequences that this failure can have is measured using a scale of 1 to 4, with 1 being not very serious and 4 being very serious. It is also indicated if the failure is critical or not for the company.
- *Failure causes.* The origin of the failure is analyzed. The goal consists on identifying the anomaly that can lead to the failure (e.g., low oil, a defective component, etc). It is important to estimate the probability P of each stoppage occurring (again on a 1 to 4 scale, with 1 being not very frequent and 4 being very frequent).
- *Actual controls.* This column indicates if, at the present time, some kind of control is carried out to avoid each cause. It also should indicate if the defect can be predicted and controlled (not the cause of the failure), that is to say, detection D (also on a 1 to 4 scale, with 1 being if the control does not always detect the cause and 4 being if it always detects).

After the first part of the FMEA table has been completed, the *risk priority number* (RPN) is calculated as the product of the three quantified variables (S, P, and D).

Using a Pareto diagram and ranking failures by RPN, the work team will analyze the causes that do not represent any threat and then choose those that could harm the process and try to eliminate them. Special attention must be paid to those effects that are considered critical, even though they may not have a high RPN.

The chosen plan of action and the employee responsible for carrying out this plan are registered in the same table used in the FMEA. Generally, after a FMEA application arises, the need for developing a preventive maintenance plan become apparent (Fig. 6.14).

In order to determine preventive maintenance intervention periods T, it is necessary to know the component-damage/wear-behavior curve.

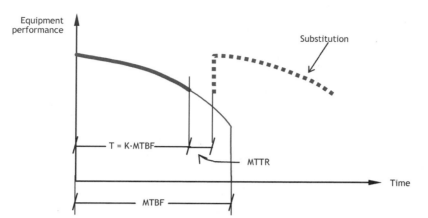

Figure 6.14. Preventive maintenance plan determination.

Information about breakdowns and the time when the breakdowns occurred is collected, and this facilitates the equipment reliability calculation in terms of mean time between failures (MTBF), as explained in the next section.

The first step is to calculate *T* based on the corrective percentage *K* that the company would like to support. Therefore, when corrective maintenance is being implemented, all the information related to the behavior of the components must be collected in order to predict future behaviors more accurately. It is also important to keep in mind that a premature or late preventive maintenance policy can be worse than corrective maintenance.

Reliability

Independent of the type of preventive maintenance plan implemented (systematic or predictive), the objectives of a maintenance plan are the same: Avoid equipment failures by increasing the reliability and effective life of the equipment at the lowest possible cost. *Reliability* is defined as the probability that a piece of equipment will work satisfactorily over a certain period of time under some specific working conditions. The following paragraphs explain these terms in depth.

Reliability is a probability. That is to say, it is not a deterministic measure of a component's useful life. This probability can be defined in many ways, although the most frequent way to define it is based on the relative frequency of breakdowns.

All production equipment should work satisfactorily. In other words, production equipment should not fail. Failure can be catastrophic or progressive. This means that failure can be triggered by an abrupt change in the component characteristic or by progressive damage. Since it is impossible to design and operate a production system that never has any breakdowns, the system should be designed and maintained to work satisfactorily for a specific period of time. This property arises because of the need for equipment elements to maintain quality standards over a reasonable period of time, since this duration cannot be infinite.

Reliability is a temporary variable, whereas quality is considered to be a momentary variable. Quality is considered momentary in the sense that it measures a product's characteristics and compares them with the product's specifications. However, reliability is the product's ability to maintain those specifications or characteristics throughout its useful life. Therefore, reliability can be defined as

$$\text{Reliability} = \text{quality} + \text{time}$$

The component or equipment life duration depends on working conditions. These conditions can be environmental (e.g., temperature or humidity) or operational (e.g., continuous starts and stops, electrical strain). Various components will not have the same reliability value under different working conditions.

The system state depends on the primary group of elements that makes it work properly, where each element has a random lifetime. Therefore, it is necessary to estimate the lifetimes of the components that wear out at a faster rate and to propose solutions so that component failures do not affect the entire system or equipment.

A good way to quantify the reliability is through the *mean time between failures* (MTBF) value:

$$\text{MTBF} = \frac{\text{operating time}}{\text{number of failures}}$$

Another important variable is *maintainability,* which is defined as the probability that if a breakdown has taken place, it must be repaired in a predetermined time following a specific repair procedure. Maintainability depends on different factors, such as

- *Machine factors,* such as accessibility or interchangeability among components

- *Organizational factors,* such as maintenance staff knowledge, documentation availability, or maintenance tasks subcontracting
- *Operative factors*, such as the ability of the staff and the clarity of working instructions

The maintainability is quantified through the *mean time to recovery* (MTTR):

$$\text{MTTR} = \frac{\text{Total time in the failed state}}{\text{number of failures}}$$

Lastly, the *statistical availability* is an average between the middle time used in the equipment and the required production time:

$$\text{Statistical availability} = \frac{\text{MTBF}}{\text{MTBF} + \text{MTTR}}$$

If the different times between breakdowns and each repair duration time are represented graphically, it is easy to understand the statistical availability concept (Fig. 6.15).

When calculating this availability, setup times and other types of idle times should not be included. Because of this interpretation, the statistical-availability concept differs from the availability concept presented in the first chapters included in the OEE definition.

Bathtub Curved. Component manufacturers usually include, among their characteristics, the failure versus time performance of their products. It is not unusual for this plot to look like a bathtub curve (Fig. 6.16).

This curve is a graphic representation of the failure rate $\lambda(t)$ that is directly related to product reliability. It can be defined as the probability that an element fails depending on its life use stage or status. The curve can be divided in three areas.

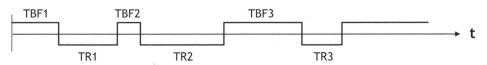

Figure 6.15. Graphic representation of time between failures and time to repair.

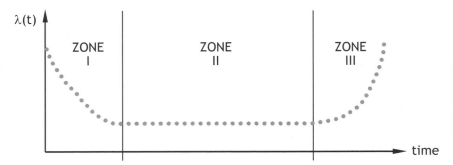

Figure 6.16. Bathtub curve example.

Zone I: Infant period. This zone coincides with the equipment setup and debugging process. It generally goes downhill because, as time moves forward, the probability of a component failure decreases. Problems in this area can be avoided by making intensive tests or by exchanging troublesome elements during an early-stage adjustment period. It is a vital stage, and maintenance workers have the responsibility of trying to shorten this period as much as they can while not affecting the component's useful life.

Zone II: Useful period. In this part of the curve, failures appear randomly. In electronic systems, there is no material wear; therefore, the curve is virtually horizontal. In mechanical systems, the curve normally has a slightly positive slope.

Zone III: Waste period. As a component or equipment reaches the end of its useful life, failures come far more quickly. In this stage, critical component replacement is strongly recommended.

P-M Analysis

The defects that take place on equipment result from two main causes: *sporadic losses* and *chronic losses*. Sporadic losses can be corrected using several tools. Some of these tools have been studied already, and others will be studied in following chapters. These losses are due to causes that can be analyzed and eliminated.

P-M analysis is responsible for eliminating chronic losses in equipment availability (Fig. 6.17), that is to say, those that are considered "natural" according to the root sources that drive them. For this reason, it is denominated P-M analysis (P = phenomenon; M = mechanism). For example, scratches on a part produced by close contact with another part with higher hardness are analyzed via P-M analysis.

In P-M analysis, the reliability that has been studied has two aspects:

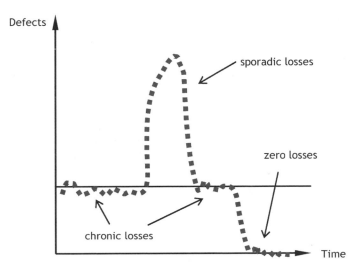

Figure 6.17. Sporadic and chronic losses.

- *Intrinsic reliability*—due to the design and production of the component
- *Operative reliability*—due to component use and the maintenance process performed on it

P-M analysis should be applied after conventional improvement and has its own implementation methodology. We will not discuss this method any further in this book.

Recently, a new tool has appeared based on statistical practices, denoted as *Six Sigma*. Six Sigma is a systematic process that has become very popular (primarily because of application results). Six Sigma is also suitable to carry out this type of study, i.e., instead of doing a P-M analysis.

Maintenance Management

A maintenance department, like any other department, should manage and control its costs properly and make sure that planned activities have been carried out. Maintenance management also should plan for future objectives for extended periods.

There are an unlimited number of indicators that can be used for maintenance department performance. Each company will decide which indicators are more suitable for it: personnel performance, hours dedicated to urgent work, repair cost, availability, etc.

In many companies, maintenance management is a difficult task because it frequently does not have upper management's support. As long as the maintenance department does not exceed its assigned budget, no one pays much attention to maintenance department activities or expenses.

Maintenance Costs. Using economic terms, maintenance management helps to control deviations in a firm's budget as well as to determine investment needs to reduce costs. Most enterprises look for ways to keep their maintenance costs low.

There are two opposing alternatives in achieving this goal (Fig. 6.18): support the cost of carrying out maintenance tasks or support the cost of not carrying out those tasks:

- *Nonmaintenance costs.* These costs derive mainly from equipment breakdowns and equipment wear-out: opportunity costs, quality costs, production personnel costs, etc.
- *Maintenance costs.* Breakdown prevention costs, anomalies detection cost, inspection resources costs, etc.

In the first part of the curve (1), maintenance investment increases the equipment availability and at the same time reduces the nonmaintenance costs. In the second part of the curve (2), an increase in avail-

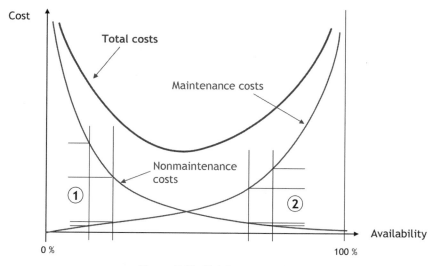

Figure 6.18. Maintenance costs.

ability supposes large investments. The lower point on the total-cost curve corresponds to the optimal availability point.

SUMMARY

This chapter has presented an overview of maintenance, a critical aspect of lean manufacturing. Maintenance planning and activities are determining factors for lean enterprises efficiency. Unfortunately, maintenance normally is perceived as a necessary evil and is not always seen as an engineering activity. This chapter has outlined some of the maintenance policies and procedures that can be used to obtain the goal of any production system: operating as efficiently as possible at the lowest cost.

RECOMMENDED READINGS

John Dixon, *Uptime: Strategies for Excellence in Maintenance Management.* Cambridge, MA: Productivity Press, 1995.

Salih O. Duffuaa, A. Raouf, and John Dixon, *Planning and Control of Maintenance Systems: Modeling and Analysis.* New York: Wiley, 1998.

JIPM: *Autonomous Maintenance for Operators.* Portland, OR: Productivity Press, 1997.

Kunio Shirose, Yoshifumi Kimura, and Mitsugu Kaneda, *P-M Analysis: An Advanced Step in TPM Implementation.* Portland, OR: Productivity Press, 1995.

François Monchy, *Teoría y práctica del mantenimiento industrial.* Barcelona: Asson, 1990.

Francisco Rey, *Hacia la Excelencia en Mantenimiento.* Madrid: TGP-Hoshin, S.L., 1996.

Seiichi Nakajima, *Introduction to TPM: Total Productive Maintenance.* Cambridge, MA: Productivity Press, 1988.

7

Equipment Efficiency: Availability, Quality, and SMED

It has become increasingly important to manufacture products economically in smaller and smaller batches. On the one hand, new management philosophies demand that product lead times (both development and then manufacturing times) are kept as small as possible. On the other hand, product customization has increased, thereby increasing the number of parts in a product family. As a result, batch sizes have been reduced and continue to shrink.

In this context, companies should be as agile and flexible as possible. Part of the required agility is to reduce machine setup times to minutes instead of hours. Unless setup time can be reduced significantly, it will be difficult to produce small batches and reduce lead time economically.

The *single-minute exchange of dies* (*SMED*) *methodology,* as it is called, is a clear, easy-to-apply methodology that has produced good results in many cases very quickly and amazing results in some other cases. The SMED methodology was developed by Shigeo Shingo in Japan from 1950 to the 1980s. With this methodology, it is possible to achieve good results without costly investments, which makes implementation in many factories an easy decision to make.

SETUP PROCESS

A *setup process* corresponds to the time required to go from the end of the last good part from one batch to when the first good part of the following batch is produced. Using this definition, the trials needed to obtain the first good product are considered part of the setup process and therefore must be studied, analyzed, and improved.

The SMED methodology is designed so that the setup process can be done in fewer than 10 minutes. During the 1960s in most automotive body shops, press changeover was consuming a large part of the available production time. It was not unusual for the setup time for a large stamping press to take more than a full day. This was Shingo's first focus, and his hope was to bring the setup times down to a few minutes.

In most cases, it is not possible to reach this objective, although it is possible to achieve reductions of around 60 percent of the original setup time. In several cases, reductions of about 90 percent have been obtained, but as a general rule, project costs become significant in obtaining these gains.

It is also necessary to point out that it will not always be necessary to reduce the machine setup process, and even an 8-hour machine setup process can be acceptable under certain circumstances. For instance, when you are replacing the tires on your personal vehicle, what does it matter if it takes an hour to change all four tires? However, in car racing (Formula 1 or NASCAR), losing 15 seconds may have very catastrophic consequences for the driver's success.

Reduction of the setup process is addressed in the just-in-time (JIT) and the 20 keys methodologies (key number 5), as shown in Fig. 7.1.

SMED implementation improves the availability rate as well as the quality rate because SMED reduces all the setup process time. This setup time includes the trial phase for qualifying the first good piece to follow setup. Therefore, startup losses included in the quality rate are produced in this trial phase (Fig. 7.2).

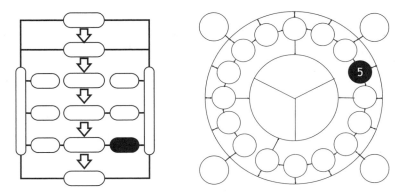

Figure 7.1. Location of setup process reduction in just-in-time and 20 keys diagrams.

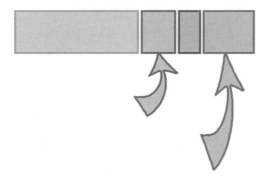

Figure 7.2. SMED improves the availability and quality rates.

THEORETICAL BASIS

Basic Steps in a Setup Process

Before embarking on an in-depth discussion of the SMED methodology, it is necessary to describe the stages that make up a general setup process. Regardless the type of the machine or equipment that is going to be studied for setup evaluation and reduction, the following classification can be used to distinguish the four typical classes of setup operations:

- Prepare, adjust, and check (new materials and tools).
- Remove old tooling and install new tooling on the machine.
- Measure, set, and calibrate (fixtures, tools and tooling).
- Produce initial parts (production trials) and adjust the machine.

These activities occur for any setup, whether they are significant (from a time and labor viewpoint) or not, and the relative proportion of time required for each type of setup operation can vary significantly. The prepare, adjust, and check operation focuses on making sure that the tools and materials that will be used for setup are available when the setup is scheduled and that idle time will not occur while the new setup material is being accumulated. The produce initial parts and adjust the machine operation depends, in most cases, on the setup specialist's know-how and therefore can be difficult to predict with precision.

Traditional Strategies to Improve the Setup Process

For early manufacturing applications, the duration of the setup process was not particularly important. Manufactures could afford to have customers waiting for their products (e.g., the next model year of a car), and production was scheduled based on manufacturing needs, so these setups hardly affected the product's price. For instance, for early automotive production, it was not unusual for a production facility to shut down for 2 weeks while machines and tools were setup for new model year production.

Today, manufacturing lot sizes have decreased. This reduction in lot size does not mean that customer demand has shrunk but rather that individual needs and expectations have increased. This points directly to the fact that manufacturing flexibility needs to increase. That is to say, if several years ago a customer requested 50,000 of a specific part, today the same quantity of parts may be requested but in increased variety and with smaller delivery quantities, which forces manufacturers to produce in smaller lots.

Unfortunately, the possible production of defective parts at setup, coupled with the increased frequency of setup, has forced manufacturers to make more products than required and in many cases store them for future orders, i.e., build to stock. For example, let us suppose that an order for 500 parts is made; the setup process takes 3 hours to get the press ready, and the defects rate is 6 percent. The machine will be scheduled to produce 530 parts to cover the possible defective parts. If the 530 parts were acceptable, it would be necessary to store 30 parts, with the related inventory costs.

In order to reduce setup process effects, companies usually use two different strategies:

- They try to make the setup as fast as possible.
- They increase the production lot size.

Skill Based Strategies. Many companies have used setup specialists to reduce changeover and setup times. These special workers are skilled in the operation of specific machines, and they are familiar with the needed tools and methods to carry out the exchange (Fig. 7.3). In some cases, a company's dependence on these specialists is so strong that the exchange has lasted more than 8 hours because the specialist did not work on the shift in which the exchange was needed.

Figure 7.3. Skilled strategy tries to make the exchange faster.

The amount of skill that a setup specialist needs for placing and removing machine elements, as well as the complexity of the setup process, seems to make improvements in the exchange more difficult on some machines. However, not all the tasks that setup specialists carry out are critical, and the specialist can get help from the machine operator, making the setup process faster and easier.

Large-Batch-Based Strategies. This strategy is based on the idea that the larger the batch size, the smaller the impact of the setup time will be on the production cost of each part. Cost per part is based on the company cost-estimation system. The system shares all company costs (direct and indirect costs, raw material costs, etc) with each product.

Regardless the cost-estimation system, there is a section in the cost per part that depends on the time to manufacture a single part as well as on its respective setup time. This time is called T_M:

$$T_M = \text{single-part production time} + \frac{\text{setup time}}{\text{batch size}}$$

If the lot size is large, the setup time effect is spread out more than if the lot size is small. As a result, T_M decreases according to batch-size increase, as shown in Table 7.1.

TABLE 7.1. Effect of Setup Time on Production Time

Batch Size	Setup Time	Production Time	Product Manufacturing Time (including part of setup time)
50	240 min	2 min	2 + 240/50 = 6.8
500	240 min	2 min	2 + 240/500 = 2.48
5000	240 min	2 min	2 + 240/5000 = 2048

T_M is proportional to the setup time. The higher the setup time, the bigger the profit, as is shown in Table 7.2.

This reasoning supports the lot size increasing. Some companies only accept orders that exceed certain lot sizes to make production with setup profitable. In many cases, a company cannot choose not to supply a product, and it is necessary to manufacture smaller lots.

In the preceding case, if setup time and production time were about the same, setup expense would not dominate the times, as shown in the tables. In that case, the time saved would be minimal (Table 7.3), and therefore, it would not make sense to search for large lots size.

Economic Lot Size Strategy. From the previous discussion, one can infer that when setup times are large, manufacture batch size also should be large. However, this policy ignores the increase in inventory cost and potential loss from products becoming obsolete.

The economic lot size is not more than a direct relationship between the inventory cost and the setup cost. This traditional formulation can be found in any book that contains inventory management techniques. In all these methods, the effect of the setup cost decreases exponentially according to the increase in batch size. Therefore, in order to calculate the economic lot size, it is supposed that the setup cost is constant; i.e., the setup time is constant.

This traditional starting hypothesis is based on a constant setup time, i.e., it is not possible to reduce the setup time. More often than not, though, setup time can be reduced (Fig. 7.4).

As setup cost decreases, the economic lot size also would decrease until the unit product lot size is reached; i.e., it would be profitable to only accept orders made up of one article—a batch size of one!

SMED METHODOLOGY

In 1950, Shingo discovered, in the Toyo Kogyo factory, that the exchange of an 800-ton press was delayed because of a missing screw.

TABLE 7.2. Effect of Large Setup Time in Production Time

Batch Size	Setup Time	Production Time	Product Manufacturing Time (including part of setup time)
50	360 min	2 min	2 + 360/50 = 9.2
500	360 min	2 min	2 + 360/500 = 2.72
5000	360 min	2 min	2 + 360/5000 = 2072

TABLE 7.3. Effect of Short Setup Time in Product Cost

Batch Size	Setup Time	Production Time	Product Manufacturing Time (including part of setup time)
50	10 min	2 min	$2 + 10/50 = 2.2$
500	10 min	2 min	$2 + 10/500 = 2.02$
5000	10 min	2 min	$2 + 10/5000 = 2002$

He understood then that two types of operations should make up the process of change:

- Operations that can be carried out with the machine running and producing parts for the previous lot. Shingo called these types of activities *external setup*.
- Operations that required the machine to be idle while they were performed, Shingo denoted those operations as *internal setup*.

In 1957, Shingo improved the SMED methodology when he realized that placing an extra table on a Mitsubishi Company machine significantly reduced the setup time. Shingo discovered that it was possible to convert some of the internal setup tasks to external setup operations.

The SMED methodology consists of four conceptual stages (Fig. 7.5), with the first corresponding to documenting the setup activities or the *current exchange study,* as it is called. Thanks to the SMED application, Shingo reduced the setup time for a screw machine manufacturer from 8 hours to 58 seconds and at the Mitsubishi Company from 24 hours to 2 minutes and 40 seconds for press setup.

Preliminary Stage

This first stage the SMED methodology consists of studying the current setup process because, simply put, "what is unknown cannot be improved" (Fig. 7.6). It is necessary to know the process, its variability, and the cause(s) of this variability.

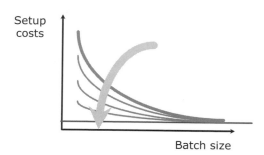

Figure 7.4. Decreasing lot size based on the reduction in setup cost.

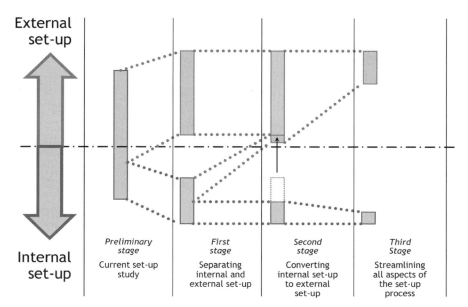

External set-up

Internal set-up

Preliminary stage	First stage	Second stage	Third Stage
Current set-up study	Separating internal and external set-up	Converting internal set-up to external set-up	Streamlining all aspects of the set-up process

Figure 7.5. SMED methodology and its impact on the setup process. Reprinted with permission from A Revolution in Manufacturing: The SMED System. English translation copyright © 1985 by Productivity Press, a division of Kraus Productivity Ltd. www.productivitypress.com.

Therefore, in this stage, it is necessary to collect data on the setup times. In some companies, setups are frequent, and it is simple to carry out several measurements. In other companies, the setups can be sporadic, and therefore, it is necessary to get as much information as possible from limited data, with only one or two setup process studies. Setup data acquisition consists of process modeling, and then, by means of the time-study process, as explained in Chap. 3, each activity

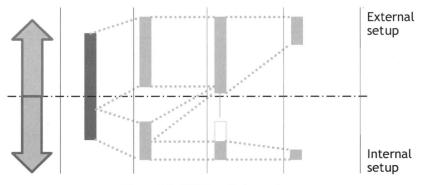

External setup

Internal setup

Figure 7.6. SMED preliminary stage.

is measured. In fact, a setup process is no more than a group of operations.

It is important when beginning such a study to clarify, mainly to the setup specialists (if the company has this kind of worker), that the goal of the SMED methodology is not to eliminate their job. A specialist always will be necessary for certain tasks. Sometimes, the opposition of setup specialist can lead to project failure, so special care should be taken to make sure that these skilled technicians do not feel threatened.

Stage 1: Separating Internal and External Setup

The first stage consists on separating the operations that should be carried out when the machine is still processing the previous lot (external setup) and those where it is necessary to carry out setup with the machine stopped (internal setup). The goal of this stage is to classify setup operations according to the given definitions of external and internal setup (Fig. 7.7). This classification takes into account the same operations and duration included in the current method, i.e., without improving any particular operation.

In addition, in this stage it is necessary to ensure that the operations defined as external setup all can be carried out while the machine is running. At first, this seems obvious, but it is always worth explaining to the worker that the necessary tools for the changeover setup and the new die should be prepared beforehand so that production time can be gained.

In practice, it is not unusual for the external activities to begin until after a batch has been completed. The main reason for this is that time to get the tools and materials necessary is not allocated to operators while they are overseeing production operations.

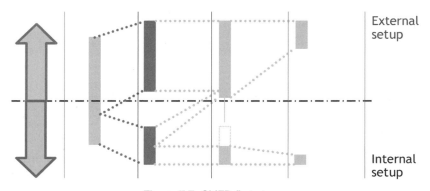

Figure 7.7. SMED first stage.

In this stage, the largest SMED cost gains can be achieved. It is not unusual to reduce the exchange time by as much as 60 percent in some cases without any capital investment.

Stage 2: Converting Internal Setup to External Setup

The setup process time reduction from the first stage can be very significant, but this is not where SMED ends (Fig. 7.8). To reduce setup time as much as possible (or economical), it is necessary to study the possibility of converting some internal setup operations into external setup operations so that they can be carried out while the machine is running.

This stage examines two important aspects of the operation:

- Reevaluate the internal setup operations to check to see if some of them were considered internal erroneously.
- Look for alternatives that allow internal setup to be carried out in whole or in part as external operations, with the machine working. For example, is it possible to screw a die to a press before placing it inside the press chamber? The answer is yes.

Logically, most of the ideas that arise in stage 2 will require an economic investment, and it will be necessary to carry out a cost justification to determine the best strategy. It is necessary to distinguish the case in which the investment is necessary despite the economics, i.e., in the case where a company could lose an important customer if delivery time cannot be reduced.

In order to decide on an alternative's viability, it is necessary not only to analyze the economics but also to study the new process or

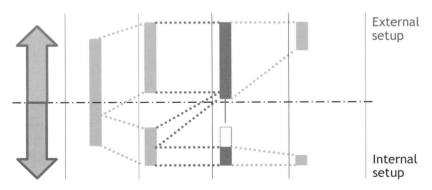

Figure 7.8. SMED second stage.

system reliability, i.e., the possible appearance of new operations (both internal and external) that increase the setup time and, of course, the benefits and possible risks of the new process.

Development of this stage can achieve, in some cases, setup process times nearing single minutes (<10 minutes).

Stage 3: Streamlining All Aspects of the Setup Process

This stage attempts to improve all the setup operations, both internal and external, reducing their duration or even, if possible, trying to eliminate some operations (Fig. 7.9). Although the SMED methodology recommends that one follow these four stages systematically, common sense sometimes can dictate that in the second stage, time and money will not be invested in operations that have not been previously optimized.

For this reason, application of the third stage usually runs parallel to the second stage, leaving a final stage 3 for the improvement in the external setup operations and a revisit of the internal activities that have not been possible to convert into external.

SMED TOOLS

First-Stage Tools

It seems logical that one should know what operations should be conducted while the machine is still processing the previous lot. Unfor-

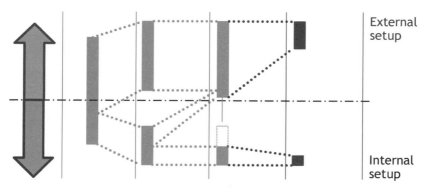

Figure 7.9. SMED third stage.

tunately, much time wasting takes place in many setup processes. For instance:

- Materials are moved to the warehouse with the machine stopped.
- Tools and dies are supplied late or incorrectly.
- Tools and dies that are not needed are taken back to the supply room before starting the machine.
- Some needed screws and tools were not collected during the setup process.
- Some nuts are just too tight when one attempts to remove them.

It is necessary to eliminate all these wastes before starting the setup. Some good questions to ask include

- What has to be done before starting the change?
- How many screws are necessary to fix the die? Of what type?
- What tools are necessary? Are they prepared properly?
- Where should the tools be placed after using them?

In order to facilitate this checking process, a group of visual controls has been developed to ensure that the needed operations are carried out before starting the setup (Fig. 7.10).

Checklist. This tool consists of a questionnaire that should be checked before each setup process. The goal of the checklist is to verify in advance that all elements that should be prepared before the machine finishes the current lot are in fact ready and available. The checklist can be universal for all product changeovers or specific for each prod-

Figure 7.10. Visual control is the most important first-stage tool.

uct. In the first case it will be placed near the machine, whereas in the second case it will be enclosed with the manufacturing order.

Check Panel. If the number of tools is small, or if the machine has its own tools, a check panel can be placed next to the machine (Fig. 7.11). Using this tool has many advantages. For example, the worker can visually check if all the necessary tools are located in the right place or if a needed tool is missing. In some check panels, the tools are silhouetted so that a missing tool is very obvious to the worker. Another strategy is to code the tools with two stickers. A sticker is placed on the machine and another is placed on the tool in order to pair the tool with the machine.

Function Checks. The checklist or the check panel do not show the die and tools status. Some plastic injection molds have material inlays that should be cleaned. If they are discovered in the trial step, cleaning of the mold can be carried out before the machine is stopped.

There are special devices for checking the molds before placing them in the machine. However, if the company does not own such a device, it might be necessary to invest in one. This possibility should be investigated and analyzed in the second SMED stage.

In other cases, it is better to inspect the mold after many pieces have been manufactured, even though the machine will be idled. Some defects will not be detected until the mold reaches the steady-state operating temperature. To check for this problem, it might be necessary to cool the mold again while inspecting the product, with a corresponding loss of production time.

Part and Tool Transportation Improvements. Part and tool transportation from the warehouse to the work area should be carried out before the exchange begins. In a traditional press exchange process,

Figure 7.11. An example of a check panel.

the worker removes the used mold, loads it in a crane, takes it to the warehouse, and loads the new mold. If the mold is heavy, the transportation movements will be slow, and time will be wasted while the machine sits idle. In order to implement this stage of the SMED methodology, it may require twice the labor, but the machine stop time decreases.

For a new changeover process, the worker would go to the warehouse while the machine is working, load the new mold, and return and leave it next to the machine ready to install. The worker then would wait until the machine finishes its work. Once the machine is finished, the worker would take out the mold and leave it next to the machine. After the worker loads up the new mold, he or she would set the machine into operation. Finally, the worker would carry the used mold to the warehouse.

This new changeover appears to take longer than the traditional method. However, according to Shingo's definition of setup time, the time during which the machine is stopped is reduced drastically. As a result, the setup time is reduced.

Second-Stage Tools

It was mentioned earlier that the second stage usually runs parallel with the third stage when an operation is optimized before convertion. However, the SMED methodology offers some recommendations that facilitate complex cases studies. For example, in this second stage, movements around the machine should not be questioned, and they must be considered as internal setup. These movements will be analyzed in the third stage. It is not that some operations will be eliminated in the second stage. However, if waste is evident, they will be eliminated. There are several methods broadly intended to enhance the setup process. Some of these methods, policies, and tools are explained briefly next.

Endless-Material Method. In some processes, reels are used to supply material to a process. When a reel is empty, it should be removed and replaced with a full one, e.g., in a rolling mill or on packing machines. The changeover times for reels potentially could be eliminated if the end of one reel is welded or tied to the beginning of the following one. The machine would work continuously. As a result, setup time would be zero.

In this case, the product made with the welded or tied seam would be scrape. However, this is offset easily by gains made from the reduction in setup time, and in many cases, the last part of the previous reel is also discarded (Fig. 7.12). In some packing lines, the previous process cannot be stopped during reel changes, and some products are scraped, which adds to productions costs and reduces quality.

Temporary Containers. Unfortunately, it is not always possible to weld or tie the material on the reels to facilitate the changeover operation. In this case, there is no other alternative but to stop the machine. Temporary containers (Fig. 7.13) do not convert the whole reel changeover operation to external setup, but they can reduce an important part of it:

- These temporary containers save setup time considerably because they eliminate the movements involved in locating and bringing in the new reel.
- If the reels allow, it might be possible to tie the previous reel with the following reel and with a simple turn carry out the exchange. This operation increases time savings.

Press-Die Preheat. In most plastic injection-molding processes, the mold has to reach a specific temperature to begin the manufacturing process. There are devices that heat such molds up before they are placed in the machine. The main concern in this case is worker safety. Manipulating hot molds represents a very dangerous task. Nevertheless, it is possible to preheat the mold to a moderate temperature, reducing the time needed for the mold to reach the working temperature once the mold is placed into the machine and thus making the process safer for the worker.

Figure 7.12. Example of an endless material method.

Figure 7.13. Example of a temporary container.

Function Standardization. A good way to convert certain mold requirements, such as height and depth adjustments, of some presses and injection-molding machines into external operations is to standardize these measures, e.g., the injector distance for an injection-molding processes (Fig. 7.14). Only the most important components for the exchange will be standardized, taking into account two main conditions:

• The setup process should be as safe as before.
• The quality of the manufactured pieces should not be adversely affected.

Sometimes, when standardizing a particular measure, it will be necessary to develop a new device. This can be the right time to add new functionalities or features to the mold, e.g., a centered guide hole.

In many cases, it will not be possible to standardize all the machine tooling owing to the large number of different tools used. Nevertheless, developing constraints and restrictions to use for the specification of

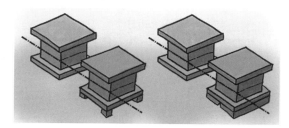

Figure 7.14. Standardization and improvement of molds.

future tooling, coupled with standardization of the most used tooling, can be very beneficial.

Tools Duplication. Sometimes it is possible to have two or more identical elements (cranes, tools, pallets, etc.) to reduce the setup time. For example, a single six pack is as easy to handle as a single can of soda. In this way, if the company has, for example, a double crane, it is possible to prepare the next mold and to extract the previous one without removing and placing the molds again. A good example is shown in the Fig. 7.15. The same scenario could be seen if a machining center had a tool changer, where the needed tools could be ready and available at all times.

Third-Stage Tools

The improvement or elimination of an operation requires reengineering of some aspect of the product or process. Reengineering can help in the analysis of several important factors. For example, is it possible to run the operation in a different way? Is this operation necessary? Is this procedure the most appropriate?

Up to this stage, external operations have not been analyzed. They simply have been distinguished, and some internal operations have been converted to external operations. However, at this point a question is in order: Will the setup specialist have enough time to organize the material and tooling and also carry out all the external operations? Although it is not considered a time-critical part of the setup (according to the definition that was given at the beginning of this section), performing external operations efficiently is always important because a setup specialist is a valuable resource.

It is necessary to study the workload of the setup specialists and to schedule the changeovers so that the specialists are not needed at the

Figure 7.15. Crane duplication.

same time on two different machines; otherwise, the work carried out to improve the setup time will be pointless.

Internal operations can be improved in different ways. In this stage it is very important to analyze in detail the movements around the machine and to determine the optimal number of workers that should take part in the setup process. Different techniques can be used to improve and to eliminate operations. Some of these techniques are presented as examples.

Streamlining External Setup

Improving Tool Storage: Indicators' Strategy. This strategy is explained in Chap. 8, dedicated to the 5S's. However, it is not necessary to undertake the complexity of a 5S methodology. The strategy here is to organize the warehouse efficiently, keeping high-use items close for easy access and organizing the tooling so that it is located and identified easily, perhaps using a code. The 5S methodology offers a common orientation when choosing coding standards so that the different sections of the company can use the same nomenclature (Fig. 7.16).

Streamlining Internal Setup

Parallel Operations. With large machines, it is necessary to carry out operations at the front of the machine as well as at the back. The worker can waste important setup time walking around the machine.

As shown in Fig. 7.17, the setup time and complexity can be reduced with the help of a second worker. If a worker needs 2 hours to carry out the exchange, two workers could take less than 1 hour when movements around the machine are eliminated (although it is also possible

Figure 7.16. Sign strategy allows ease of use in tool storage.

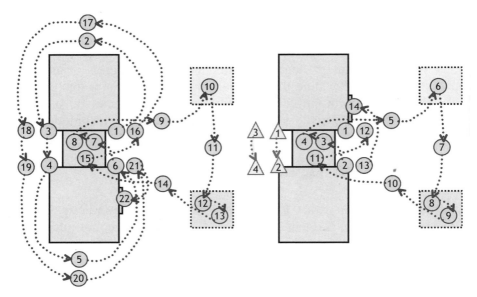

Figure 7.17. Comparison between one or two workers in a setup process.

that they could use more time based on task sharing and the operations sequence).

It is convenient to use a task map to indicate what operations will be carried out by each worker, starting with the initial situation shown in Table 7.4.

When two workers execute the exchange in a parallel fashion, the exchange procedure is presented in a table (Table 7.5). The table shows the task sharing and those tasks that can be carried out simultaneously, as well as the times where the workers should be waiting.

This representation allows the lean thinker to discover which are the most important tasks that should be improved or which are the tasks that the other worker can carry out in order to make the first worker's job easier (Table 7.6).

Worker safety once again is a priority in this type of synchronized work. There are special devices that decrease the probability of risks to workers, such as safety mats that stop the machine when they are activated. Safety mechanisms that halt the machine until some signal can be provided, e.g., confirmation buttons, etc., can save accidents and injury.

One-Motion Method. Some setup processes allow machines or people to perform more than one task simultaneously, e.g., performing electric and hydraulic connections on a refrigeration circuit while the mold is

TABLE 7.4. Initial Situation

No.	Procedure
1	Remove front bolts
2	Movement to machine back side
3	Remove back bolts
4	Remove back wiring
5	Movement to machine front side
6	Remove front wiring
7	Put used die up
8	Lift used die
9	Move used die next to machine
10	Remove used die
11	Move crane to new die
12	Put new die up
13	Lift new die
14	Move new die to machine
15	Get new die down
16	Fix front bolts
17	Movement to machine back side
18	Fix back bolts
19	Fix back wiring
20	Movement to machine front side
21	Fix front wiring
22	Fix new machine parameters

TABLE 7.5. Task Sharing and Simultaneous Operations

No.	First Worker Procedure	No.	Second Worker Procedure
1	Remove front bolts	1	Remove back bolts
2	Remove front wiring	2	Remove back wiring
3	Put used die up		
4	Lift used die		
5	Move used die next to machine		
6	Remove used die		
7	Move crane to new die		
8	Put new die up		
9	Lift new die		
10	Move new die to machine		
11	Get new die down		
12	Fix front bolts	3	Fix back bolts
13	Fix front wiring	4	Fix back wiring
14	Fix new machine parameters		

TABLE 7.6.   Improvement in Task Sharing

No.	First Worker Procedure	No.	Second Worker Procedure
1	Remove front bolts	1	Remove back bolts
2	Remove front wiring	2	Remove back wiring
3	Put used die up	3	Movement to machine front side
4	Lift used die	4	Fix new machine parameters
5	Move used die next to machine	5	Movement to machine back side
6	Remove used die		
7	Move crane to new die		
8	Put new die up		
9	Lift new die		
10	Move new die to machine		
11	Get new die down		
12	Fix front bolts	3	Fix back bolts
13	Fix front wiring	4	Fix back wiring

being slowly fit into position. In the case where these connections cannot be carried out simultaneously, similar connections can be grouped onto a device that allows an operator to connect all the connections even faster.

Functional Clamps. Functional clamps are devices that are used to passively hold an object in a fixed position with the minimum effort. Setup tasks such as turning nuts in and out or tying and untying a component can be eliminated.

The SMED methodology seeks to eliminate the use of screws and nuts as fixing elements. Shingo found that of the entire screw body (thread), the only thread that carries the press function is the last one. Therefore, all the screws should have only one thread. In this case, the screw would be considered as a functional clamp.

From a technical point of view, a one-thread screw or bolt is not viable, but in many cases the number of screw threads can be decreased. There are a large number of functional clamps that, although many of them are used with screws, facilitate installing, fixing, and removing dies.

- *Pear-shaped holes.* In many tops or surfaces with a great number of screws it is not necessary to unscrew all the bolts until the end. Using pear-shaped holes (Fig. 7.18), it is enough to unscrew them only until the top can be turned.
- *U-shaped washers rings.* These are similar to the clamp, but they are used in a large number of cases because they are helpful in any joint between a screw and a nut.

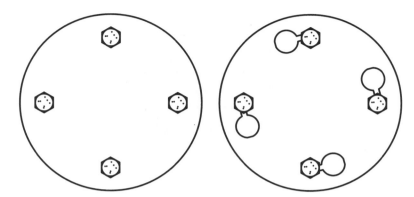

Figure 7.18. Pear-shaped holes are functional clamps.

- *C-shaped washers.* These washers are used in difficult access situations when there is a risk of loosing a U-shaped washer into the machine.
- *Guttered thread.* This is an approach to the ideal screw with only one thread. The guttered thread allows a worker to fix an element with just one-third turn.
- *Single-movement method.* Axels that do not turn at high speeds can be held with simple elements that can be let free by turning a lever.
- *Reduce tools variety.* This can be done by using all the same screws so that the same tool can be used at all times or having simple earflap bolts.

Eliminating Testing Procedures (Trials) and Adjustments. Adjustments and trials with many changes easily can represent 50 percent of the total changeover time, i.e., until the first good piece is obtained. Trials and adjustments should be viewed carefully so that they are not only decreased but also eliminated. Whenever possible, a set point should be used instead of an adjustment (Fig. 7.19).

In many companies, the settings and positions for the next tooling can be used without applying the trial-and-error method. If adjustments are necessary, the adjusting procedures must be written, and the machine parameter values should be perfectly specified. In this way, if trials and adjustments must be performed, they will only be those that are extremely necessary. A new technique has been developed in which the adjustments sometimes are eliminated. It is called *one-touch exchange of die* (OTED).

Figure 7.19. Trials and adjustments should be eliminated.

Process Automation. A fully automated process is the last resource after successful implementation of the preceding methods. It will be effective if and only if the operation on which the automation will be applied is already optimal. Generally, process automation supposes high investment costs (Fig. 7.20). Regardless of the expense of implementation, in some cases it is the best alternative.

Zero Changeover

With SMED implementation, amazing time reductions can be achieved. However, it is the small improvements that can be achieved with SMED that are much more significant.

When Shingo died in 1990, Sekine and Arai continued his work, trying to go further than Shingo did in terms of time reduction. They atried to achieve setup times of less than 1 minute (seconds). In order to achieve this goal, they create a strategy called *zero changeover.*

Unfortunately, the only way to achieve setup times in the seconds is to automate the exchange process, which represents a large investment.

Figure 7.20. Automation is the last method to apply.

Many of the improvements achieved, came from ideas developed by Shingo.

SMED EFFECTS AND BENEFITS

SMED provides several benefits. In the end, all these time reductions are translated into money savings, although there are other aspects, such as safety improvements, that are difficult to quantify economically.

Easier Setup Process

The reduction in the number and complexity of operations that the SMED methodology provides leads to the implementation of new changeover procedures. As a result, the setup process is simplified, and it becomes easier to carry out.

Thanks to the SMED, many of the operations can be carried out by most employees (fewer skills are required). Therefore, the setup specialist and the worker collaborate in the setup process.

Increased safety also results from the improved changeover simplicity. To reduce the setup time, all operations have to be analyzed in detail, which at the same time eliminates risky situations.

The new setup procedure can eliminate defective parts in the manufacturing process, making sure that when all the setup steps are performed properly, the machine is able to produce correct products from the first part.

On-Hand Stock Production

If the setup time decreases, manufacturing batch size can be decreased as well. Therefore, it will not be necessary to make large manufacturing orders, and as a consequence, work-in-process will decrease. If the work-in-process decreases, the mean time of material flow will decrease, as well as the lead time, because it is directly related to the time of material flow.

Workplace Task Simplification

Tool coding and a clean and upstanding machine environment, among many others, are strategies that help to simplified the workplace. After

SMED implementation, it is easier to locate tools, dies, and parts in a short period of time.

Productivity and Flexibility

The benefits of SMED implementation are many and important, but the main benefits are centered on two key concepts: an increase in productivity and an increase in flexibility (Fig. 7.21).

Sometimes the SMED methodology is applied to decrease a machine's load in order to increase the productive period. Although the productivity increases, thanks to the SMED methodology, SMED's principal benefit is to increase flexibility, as will be shown next. If a company needs to improve a machine's flexibility and must decide between buying a new machine and spending the same amount of money on improving the setup time, by means of a SMED implementation, the decision is clear, as illustrated in Fig. 7.22.

It can be concluded that if a company looks only for a productivity increase, then SMED is one among a number of other methods, e.g., buying another machine or eliminating idle times. However, if the company is looking for an increase in flexibility, SMED is the only solution because buying a new machine does not provide an increase in flexibility.

Economic Benefits

The economic benefits derived from SMED implementation are not always the same and depend on the machine arrangement to which the SMED is applied.

- In some cases the machine on which the methodology is applied is saturated. If the objective of the SMED is to liberate the machine

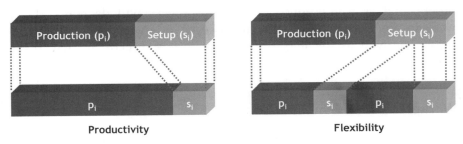

Figure 7.21. The main SMED benefits are improvements in productivity and flexibility.

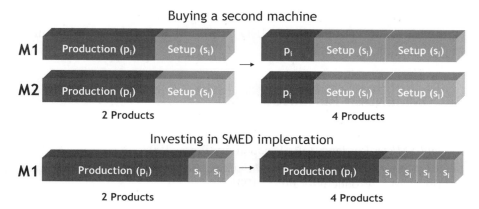

Figure 7.22. A second machine does not improve flexibility if the number of exchanges increases.

from its load time to increase machine availability, the benefit takes place because of the economic margin in the sales increment.

- If the machine is not saturated, and the number of changeovers is not important, the time needed to carry out a production order will decrease. If the machine's workers can be assigned to other sections, the economic benefit results from the saved labor costs.

These two examples demonstrate that in each case the economics of SMED can vary.

A flexibility increase is very difficult to measure economically, and it is necessary to relate it to the stock level (decrease in stock) or to other quantifiable benefits, e.g, the satisfaction of customers who receive their product orders in a shorter time. Nevertheless, the economic justification of a SMED study that looks for a flexibility improvement is always complicated. In this case it is also possible to present qualitative benefits derived from SMED.

SUMMARY

This chapter has presented a basic but helpful methodology to reduce the setup time in a machine: the SMED methodology. The SMED methodology proposes arranging the tools needed in the setup process before the machine finishes the preceding lot. The main benefit of setup time reduction is on increase in the flexibility of production. Flexibility is a key concept in the lean manufacturing strategy. Nevertheless, the economic justification of a SMED project that focuses on flexibility improvements is quite difficult.

RECOMMENDED READINGS

Kenichi Sekine and Keisuke Arai, *Kaizen for Quick Changeover: Going beyond SMED*. Portland, OR: Productivity Press, 1992.

Shigeo Shingo, *A Revolution in Manufacturing: The SMED System*. Cambridge, MA: Productivity Press, 1985.

8

Environmental Improvements and the 5S Methodology

The third area that will be analyzed in this book is the work environment and how it can be improved. In the last decade, the number of implementation projects based on a methodology known as the *5S methodology* has increased significantly. The name of this methodology corresponds to the initial letters of five Japanese words (also five English words) that are based on *sort, organize,* and *clean.*

However, the main objective of the 5S tool is to educate workers and encourage an attitude that supports worker habits. These habits will allow workers to maintain the work environment in an orderly (sorted, organized, and clean) manner with little effort. The ideas used in this methodology are simple, and most of them are based on common sense. However, in most companies, these procedures of organization and cleaning are not adhered to as well as they should be.

A CLEAN AND ORGANIZED WORKSPACE

Before getting started on production improvements, it is necessary to have a clean and organized workspace. Hiroyuki Hirano developed a methodology that leads to "working with essential elements," as well as to an organized and clean workspace (Fig. 8.1).

The 5S methodology has brought a new awareness of, as well as a respect for, organization and cleaning into the company environment. This attitude is not based on posters or stereotyped slogans hung on

Figure 8.1. The 5S methodology allows one to avoid the situation on the left and maintain the one on the right.

factory walls. The 5S methodology is based on inculcating, through 5S tools, good habits that will ease future implementation of improvement tools.

For this reason, in the just-in-time (JIT) philosophy, the 5S tool occupies the first place in the diagram (Fig. 8.2). The 20 keys methodology also separates this key from the rest and locates it outside the circle (key number 1).

The 5S methodology is identified with five Japanese words:

- *First pillar: sort* (*seiri*). When applying this pillar, elements that are necessary and those that are not should be differentiated.
- *Second pillar: set in order* (*seiton*). The objective of this pillar is to be able to organize the necessary elements so that anyone can

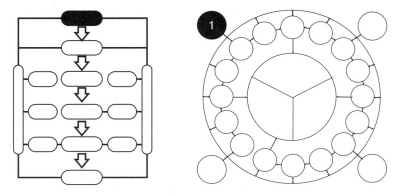

Figure 8.2. Location of the 5S methodology in just-in-time and 20 keys diagrams.

find them, use them, and return them to the same place after their use.

- *Third pillar: shine (seiso).* This pillar focuses on the necessary tasks to clean the working area.
- *Fourth pillar: standardized or visual control (seiketsu).* This pillar keeps active the three previously listed pillars. After the effort that these three pillars supposes, one cannot let the work already done go to waste. In addition, detecting anomalies in the process becomes easier.
- *Fifth pillar: sustain (shitsuke).* These new working procedures need to be enforced until they become habit.

5S IMPLEMENTATION METHODOLOGY

Getting Started

Before starting with a 5S implementation, there is an initial step that is basic to success of this methodology. This initial step consists of convincing management that even though at first the 5S methodology presupposes taking time away from production to implement and that new tasks (shine and sustain) will remain forever, in the long run this methodology will increase productivity. It is also necessary to

1. *Prepare didactic material.*—to explain to all the workers the importance of the 5S methodology and the basic knowledge that all of them should have about this methodology.
2. *Choose a pilot area.*—where spectacular results can be achieved in a short period of time or where workers are more motivated with the project. This area should not be very big. The objective of the pilot experience is to create and spread out expectations into other areas. If it is possible, avoid choosing an office as the pilot area because it is difficult to extrapolate the achieved results to the production area.
3. *Design a working plan.* There are some questions that should be answered before starting with the basic formation: When will the team work? Who will be on the team? Will the work be remunerated? Will this project change the incentive system?
4. *Prepare the training plan and complete the methodology for the working team.* The 5S methodology proposes for each S the following plan of action: (1) Train workers on each S, (2) put it into

practice, and finally (3), share the experience. In addition, the working team should have a panel, a digital camera, and other small but important devices such as portfolios, pens, etc.

Once the results of implementation of the 5S methodology in the pilot area are known, the methodology can be extended to other areas, where members of the first working group typically become facilitators for the new groups.

It is not unusual for some improvements that were implemented in the pilot area to solve problems in other areas. If this is the case, and especially if the implementation will take time, it is convenient to carry out those small improvements in these "other areas" to avoid stress among the workers.

However, the same situation can have a negative effect because in these areas workers have not been trained or traveled the road to discover the need for the 5S methodology and also because the 5S team can feel underappreciated.

Training programs concerning the reasoning behind and the scope of the 5S implementation are considered a basic step. Nevertheless, it is very important to count on the support of management. Without this support, the 5S implementation will fail because in some cases it will be necessary to impose certain procedures, and only a person with higher authority will be able to impose them.

Finally, it is important not to set very demanding goals in a 5S project. If the intention is to develop many improvements, and some of them fail, the whole project can be seen as a failure. For example, the 5S methodology may intend to improve worker results or methods. However, this represents a process improvement and should be viewed as an independent project. As a result, it should be separated from the 5S implementation process. That is to say, it should not appear in the 5S panel or be discussed in the meetings.

Common Steps in the Five Pillars

The 5S plan cannot be based on intuition or implemented in pieces. A simple methodology that has been successful in many companies is described by 5S author Hiroyuki Hirano. In addition, the 5S methodology can be studied in specialized books, and all of them recommend the same steps before implementation of this methodology. In Europe, there is also an extended 5S methodology that was developed by Bekaert Consulting.

In both approaches, the methodology is carried out in a systematic way. The procedure to follow for each approach is as follows:

- The team is formed.
- The tools of each pillar are used (these tools are explained at the end of this chapter).
- Some indicators that allow verification of the state of the implementation are established.

First Pillar: Sort

In any company, numerous disorganization symptoms can be found:

- Obsolete or retired equipment that is frequently placed in low-traffic areas near walls. The equipment remains there for a long time, and if it does not interfere with normal plant operations, it never get stored or disposed of. Another disorganization symptom is the lack of a specific area for work-in-process (WIP).
- In some companies, workers have to move around machines, objects, and parts in order to go from one area to another. The problem becomes more serious if the workers are manipulating parts with forklifts.
- Another cause of disorganization is the accumulation of obsolete pieces from machines that are no longer used or products that are no longer manufactured.

People usually place a high sentimental value on familiar objects: tools, machines, documents, etc. As a consequence, workers cannot decide which objects are necessary and which are not. The usefulness for almost all the objects can be found easily.

The message that the first pillar tries to get across is very strong: It is necessary to get rid of all the unnecessary objects! In order to achieve this goal, objects can be grouped into three categories:

- Those that are used frequently.
- Those that probably will be used.
- Those that never will be used.

Elements belonging to one of the last two categories should be taken out of the work area. Temporarily, those from the group of elements that "probably will be used" can be stored in a special area that should be prepared, to avoid throwing them out.

Second Pillar: Set in Order

The second pillar for 5S implementation requires that the first pillar be completed because it does not make any sense to set in order unnecessary objects. The main goal of the second pillar is to cut the time required for (or completely eliminate) material searches as well as to facilitate the movement of objects through the factory. Some of the most common wastes are an inability to find a tool, to have a drawer with mixed and unordered components, to have unnecessary doors that open and potentially endanger somebody that is near, etc.

Hiroyuki proposes not only to organize but also to take advantage of these improvements in order to facilitate manufacturing processes (Fig. 8.3).

In order to carry out these improvements, methods and time-study tools, already studied in previous chapters, can be applied. However, this process can suppose, in some cases, an important investment of time to achieve the improvement objectives. Therefore, it is recommended that at first you only carry out those actions that are urgently needed and then assign the needs discovered to other independent projects.

Third Pillar: Shine

This pillar attempts to implement factory cleanliness. It simply states, "Get rid of dust, grease, filings, or oil from the working places." In other words, maintain the factory clean and swept at all times.

In a factory, as in a house, when cleaning is not practiced routinely, the windows are dirty and less light gets in, the corridors seem darker,

Figure 8.3. Organization sometimes implies task improvement.

and there exists a risk of injury because of such conditions. The lack of cleaning supposes risks (Fig. 8.4). For example, oil on the floor produces a slick surface, a nail can pierce shoes, etc.

Finally, the lack of cleaning also can facilitate equipment breakdowns. For example, dirt can hide the oil-level indicator, which could lead to machine failure and reduce its lifetime.

The Labor Risks Prevention Law is a mandatory law for companies. It has begun to make companies more aware of the importance of maintaining a clean workplace.

Owing to the effort required for cleaning, many companies now outsource these services to specialized cleaning companies. Such companies usually take care of general cleaning, but in most cases they do not clean machines and tools (very important cleaning areas).

Regardless of the cleaning method selected, it cannot be limited to "spring cleaning." It should be a constant attitude and, if possible, a daily task. That is, cleaning should become a habit.

Once a cleaning method is implemented, inspection elements also can be added to it or, more appropriately, carried out simultaneously. At this point, the company is able to implement some autonomous maintenance procedures.

It is important to keep in mind that the starting point for this improvement was the 5S implementation, not a maintenance project. Therefore, it is recommended that these maintenance inspection elements not be included in a 5S project, and a separate maintenance project should be conducted. This point shows, once again, the close relationship between 5S and other improvement activities.

Fourth Pillar: Standardize

The fourth pillar does not pursue a concrete objective like the three previous pillars do. The standardize pillar is achieved when the three

Figure 8.4. The lack of cleaning can result in a risk of sliding.

previous pillars are implemented and running routinely (Fig. 8.5), basically making these activities easier for workers to maintain (reinforcing the habit). There are some tools that make standardization of the sort process, order, and cleaning operations easier to transform into habits.

The standardize pillar adds the word *preventive* to each one of the three pillars so that now the objective is to avoid the need to repeat the initial effort of each pillar. For example, if workers step on oil, then everywhere they go will become a mess. As a result, it is necessary to clean regularly. Preventive cleaning looks to eliminate the source (the oil source in this case) of the problem. However, the proper approach would be to start an equipment maintenance study that potentially would eliminate the oil source by means of a corrective or preventive maintenance task.

Figure 8.5. Visual control allows us to detect anomalies with just a quick look.

To achieve the fourth pillar goal, which involves transforming the preceding three pillars into habit, it is necessary to assign responsibility to the workers. In other words, it is necessary to integrate some cleaning and ordering tasks in the worker's regular work (daily routine) and, if necessary, to watch over their execution by means of *5S audits*.

It is usually possible to detect anomalies in a visual way, and the action to correct the anomalies should be immediate (what is known as the "5 minutes of 5S"). For this reason, this pillar is also known as *visual control* (Fig. 8.5).

Fifth Pillar: Sustain

In many aspects of our daily life, discipline helps us to achieve our objectives. If we decide to lose weight via a diet, we start with great enthusiasm, but then we will lose motivation as time goes by. Therefore, discipline is important to achieve any goal (Fig. 8.6). In order to maintain motivation, it can be useful to know beforehand the benefits that will come from the assigned tasks.

In a company without discipline, the working space can become eroded in a heartbeat, and unnecessary objects will accumulate again. We recommend that it is made very clear from the beginning of 5S implementation that if discipline does not exists, these goals will never be achieved.

An important point to make is that if the workplace returns to its initial state of "disorganization" after implementation of the 5S meth-

Figure 8.6. Knowing the benefits allows us to maintain our motivation.

odology, it will be more difficult to implement 5S again. It should be done properly right from the start.

We impose discipline at the beginning of 5S implementation with the hope that later on, the 5S approach will turn into habit. Few people have the authority to impose tasks on workers in a company (company mangers and section heads). It is also important to count on the support of workers who lead worker groups (shift leaders or other similarly titled people). Therefore, discipline is the pillar that sustains the four previous pillars because it leads to good work habits in the workforce.

IMPLEMENTATION OF THE 5S METHODOLOGY IN OFFICES

Many corporate offices are more disorganized than the production plants of our preceding discussion. Although the 5S methodology describes numerous improvement examples in the production environment, it is not limited to the shop floor. We should not forget other higher-expense areas of the company, such as offices, warehouses, common areas, etc. Table drawers accumulate useless objects, especially in the middle drawer! The same scenario happens with closets and paper piles. In implementing the "organize" pillar, we recommend avoiding generic words such as *others, several,* and similar words. They can be confusing and meaningless to a third party.

The last point to consider in offices is office supplies. There should be a procedure to restock office supplies. In some companies, printing supplies such as ink and paper are stored in such a way that after a certain level is reached, an order goes out to the supplier (for a specific reorder level). Supplies typically are low-value items that seldom become obsolete, so we recommend not going to extremes in setting up a system.

Applying 5S to Computers

The human brain carries out a technique called *self-erasing*—forgetting those things that they are not used frequently. Using this technique on the computers in each office could lead to significant improvements (Fig. 8.7).

In order to be efficient at work, it is vital to maintain computers in an organized and clean (lean) manner. The fear of loosing information should not stop workers from updating and cleaning their computers and other information system equipment.

Figure 8.7. Computers are office machines that also need 5S implementation.

Folder trees can be designed so that information can be organized in a standard way. Information then can be "backed up" only for data files rather than the entire computer memory. The same recommendation can be applied to the e-mail inbox.

5S TOOLS

Red-Tagging Strategy

This is a simple and visual method to separate necessary elements from unnecessary ones, and it is also used in the first pillar (sort). This technique consists of assigning a red card to elements that are not used or whose use is unlikely.

There are many reasons to use red (Fig. 8.8). It is a bright color; it is the color that is used in traffic lights to indicate a stop; and in Japanese, the word *red* is used to denote "dirty" (although in Russian it means "beautiful"). The reason for placing red cards, instead of removing the elements directly, is that it allows a picture to be taken

Figure 8.8. The red color is used in wrong situations.

that will be placed in the panel in order to illustrate the evolution of the workspace (before and after).

When carrying out the red card campaign, it is not necessary to be embarrassed if the entire plant becomes red. It is recommended to put red cards on elements that will not be used in the next month, although at first red cards will be placed only on elements that are completely unnecessary.

Red cards should never be placed on people's tables. In a manufacturing company, red cards should be placed on equipment and on obsolete inventory.

If somebody asks a worker if an object is necessary, the answer always will be affirmative. Therefore, when in doubt, it is recommended to place a red card on the object.

Using a second color, yellow, for example, for a card that implies "may be useful" should be avoided. In a company where this methodology was used, the 5S team opted to place yellow cards for "may be useful" objects, and no red cards were placed. Therefore, only red cards are allowed.

Sign Strategy

In cities, on highways, in clothing stores, and virtually everywhere in our daily life, signs that facilitate our ability to locate a place or an object can be found easily. A simple example is the sign for a men's or women's restroom. However, in factories, posters and signs are considered decorative elements. It sometimes seems that they do not add any value to the plant, and they only generate wasted time that could be dedicated to other productive tasks. The signs strategy is used in the second pillar (set in order).

Signs can be placed on machines (Fig. 8.9), as well as in various areas of the plant, but the most important are those that make reference to stock and tools because they facilitate the searching tasks.

Using posters and signs has many advantages. One of the most important advantages is for new hires to the company so that they can easily find their way around and learn what each factory section does while learning the language of the other production workers. It is recommended that signs be placed and displayed in factories in a similar way to how they are placed in cities. In some warehouses, this methodology is applied to the layout and labeling of the aisles (streets), where the even numbers correspond to one side of the plant and the odd numbers correspond to the other side (Fig. 8.10).

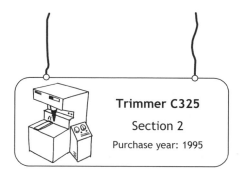

Figure 8.9. Signs facilitate production processes.

The *element sign* makes it easy to see if an element is placed correctly. For example, in a private parking area, it is hard to ensure that vehicles are located correctly if the parking identifier does not correspond with some characteristic of the car. For example, the registration number (license plate) should coincide with number of the car's slot (Fig. 8.11). Products and raw material do not have registration numbers, but they can be coded (Fig. 8.12).

Besides product codification, certain other codes, such as product quantity signs, can ease inventory management. Product quantity signs help workers to manage inventory levels in a visual way because the minimum and maximum quantities of each articles can be preassigned. In order to implement product quantity signs, a study of the warehouse may be needed. The 5S tools do not provide any methodology to carry out this type of study. Therefore, it is convenient to carry out this procedure only looking for volume capacities and, if needed, carry out a more exhaustive warehouse study later.

Figure 8.10. Warehouse signs can be similar to those used for streets.

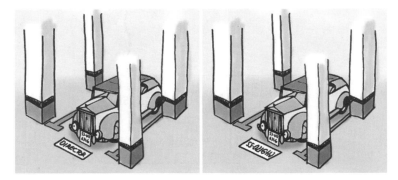

Figure 8.11. The sign in the left figure does not ensure that the correct car is parked in the slot.

Painting Strategy

The painting strategy focuses primarily on floors and walls. Its main goal is, in the first place, to separate walking areas from working areas. It is also a tool used in the second pillar (set in order).

Thanks to the painting strategy, materials handling becomes easier because the working areas are delimited. Also, areas where workers should not enter are clearly identified.

It is strongly recommended that when painting the plant along with the corridors and corners, the painting design should be as straight as possible. Avoiding corners and twisted shapes like those on the right side of Fig. 8.13. Moreover, when painting the plant, bright colors should be used for the lines (yellow, orange, or white). Green and blue should be used for working areas.

Areas where doors open also should be painted so that fork trucks are not parked within them, thereby preventing their use or potentially

Figure 8.12. Product codification facilitates product arrangement.

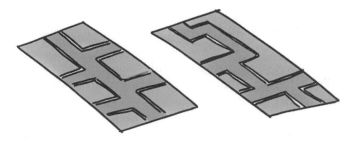

Figure 8.13. Factory painting should be as straightforward as possible.

causing accidents. Areas designated for work-in-process again should be painted. It is recommended to paint only the corners of these areas; otherwise, the factory floor can end up looking like a sports center. Lastly, areas of high risk should be identified with tiger marks (yellow and black), e.g., a hose across a corridor or the action area of a robot.

The ordering of templates and tools was analyzed in detail when the SMED methodology was studied. It quite possible that it is not necessary to implement the SMED system for all equipment, but rather to implement the 5S methodology for simple systems. In this case, the recommended strategies are the silhouette method or color coding (Fig. 8.14). In a SMED implementation, the 5S methodology offers uniformity in the corridors and shelving coding for all company areas.

Preventive Order

The preventive order has the principal objective of avoiding a return to the disordered scenario. Preventive ordering looks for challenging tasks that might easily fall back into disorder. For example, in the case of hand tools, three techniques that prevent chaos are

Figure 8.14. Color coding facilitates tools ordering.

- *Suspension.* Tools are suspended from above with a pulley and a spring. This technique is used frequently for pneumatic tools, where the weight of the tool also can be offset.
- *Incorporation.* Some gauges or rules can be glued to the machine to facilitate the adjustment of some measures, with target marks specifically identified.
- *Use elimination.* Many times combining tool functions or eliminating tools by using a standard size nut or bolt can be achieved with minimal investment. It also may be possible to change the fixing device (snap versus a screw).

Preventive Shine

The objective of preventive shine is to anticipate sources of dirt or foreign matter before they occur. That is to say, the objective is to avoid the need for cleaning (Fig. 8.15).

It is necessary to develop an in-house awareness of use of safety devices such as guards. They are required by the Labor Risks Prevention Law (Europe) and the Occupation Health and Safety Administration (USA).

Promotional Tools

In Japan, the team responsible for implementation of the 5S methodology has a clear vision from the very beginning of the need for discipline and the importance of transforming the new working procedures into habit. In order to promote this idea, the team uses some promotional tools in the project that have the goal of motivating factory staff.

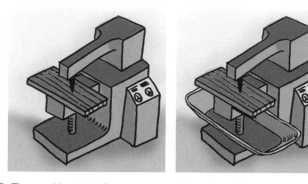

Figure 8.15. The machine tray allows one to avoid cleaning the machine's surrounding area.

The main promotional tool is the enthusiastic way in which the team carries out the 5S implementation project (Fig. 8.16). This reduces the disruption that implementing the process may bring to shop floor workers.

There are a great number of promotional tools. Some of them are listed below:

- *Slogans.* In Japan, posters are used in the 5S implementation. These posters contain short sayings (called *haiku*). For example, "Even factories feel good taking a bath."
- *5S news bulletins.* Reports that are published in a very simple way but that present ideas, improvements, and pictures also can be effective.
- *Pictures panels.* The placement of a panel with pictures of the workspace evolution reflecting the steps that have been taken can be highly motivating for the workers.
- *5S competition.* The Japanese carry out competitions among different sections over a two- or three-month period every year. During these months, called *5S months,* seminars about the importance of 5S are given.
- *Emblems and buttons.* Buttons or patches are prizes that are granted to workers from the sections with the most points or those that receive a specific level that the 5S auditors compute.
- *Pocket manuals.* The publication of small manuals that illustrate ideas to implement the 5S methodology in the workspace is very useful. These pocket manuals are distributed to all workers and help to diffuse resistance to improvements among different departments.

Figure 8.16. Enthusiasm is the main promotional tool.

5S BENEFITS AND EFFECTS

The benefits obtained when applying and maintaining the 5S methodology go beyond just improving the workspace environment. It is possible to relate the 5S implementation to other improvement tools, as will be shown next.

The 5S methodology, independent of SMED implementation, can improve tool and die location in a warehouse and, as a result, can indirectly decrease equipment setup times. This can increase the variety of products that can be produced economically and, therefore, equipment flexibility.

Thanks to the 5S methodology, the probability of assembly errors (wrong parts) is decreased by having coded stock cases with well-identified parts. In the case of automatic equipment, performance normally is more efficient when the equipment is clean. Thanks to the 5S methodology, product quality is higher.

A large portion of the equipment waste (idle time) is related to component and part searches, which are eliminated thanks to 5S. Therefore, idle time and cost decrease, as well as the cost of the product.

Clean and well-delimited floor spaces avoid the risk of sliding or falling. Properly organized warehouses avoid clutter and its associated risks. The Labor Risks Prevention Law forces management to carry out many of the activities already conducted in the 5S methodology.

Cleaning can increase equipment availability by reducing the time required to carry out maintenance tasks. Therefore, 5S implementation has proven to have a significant impact on other areas in the company. SMED, TPM, quality, labor risk prevention—almost all these tools are made easier thanks to 5S. One should not make the mistake of thinking that the 5S methodology solves all these problems because it is not true. The habits resulting from 5S implementation will ease the future implementation of other improvement tools.

Habit is the main 5S benefit, and it allows us to understand why, in the just-in-time and 20 keys philosophies, 5S implementation is one of the first and higher-priority points. Some companies are aware of the difficulty of achieving the benefits that the 5S habits provide. Therefore, it might be a good idea to get started with the 3S strategies (organization, order, and cleaning) first. Another advantage of the 5S methodology is that it is possible to put it into practice very quickly (Fig. 8.17). It is possible to achieve order and good cleaning habits before beginning any production in a company.

The 5S implementation program has a deeply personal reward. The main personal reward of 5S implementation is to work in a pleasant

Figure 8.17. 5S at home is the best practice area.

place. It is said that in a third-class workspace, people leave garbage, and nobody stops to pick it up; in a second-class workspace, people leave garbage, but somebody stops to pick it up; finally, in first-class workspace, nobody leaves garbage, and, everybody is willing to pick it up.

SUMMARY

This chapter, focused on one of the extended tools of the lean manufacturing philosophy, gathers together some common actions explained in previous chapters. Nevertheless, the 5S methodology has an objective that cannot be reached by means of any other improvement methodologies: Educate and maintain an attitude in order to support workers' habits. The 5S methodology builds a culture in the company that eases the implementation of the rest of the improvement methodologies and techniques.

RECOMMENDED READINGS

Hiroyuki Hirano, *5S for Operators: 5 Pillars of the Visual Workplace.* Portland, OR: Productivity Press, 1996.

Hiroyuki Hirano and Bruce Talbot, *5 Pillars of the Visual Workplace: The Sourcebook for 5S Implementation.* New York, NY: Productivity Press, 1995.

9

Other Improvement Keys

In previous chapters, different improvement tools that can be used to solve many production problems were discussed, illustrated, and analyzed. These tools are included in the book, *20 Keys to Workplace Improvement,* but not all 20 keys have been explained. The rest of the keys can be grouped in four categories:

- Human resources–related keys
- Efficient materials use–related keys
- Visual control–related keys
- Technology–related keys

In this chapter, all these tools will be described briefly. To analyze them in depth, it is necessary to consult Kobayashi's book. We also include a discussion of the *jidoka* (autonomotation) tool within the technology keys group.

HUMAN RESOURCES–RELATED KEYS

Rationalizing the System

All the improvement tools described so far cannot be applied effectively without the cooperation of all the workers from all the departments or organizational units of a production system. A frequent response when one requests help from certain workers is, "That it is not my problem or my job." Therefore, the obligations and responsibilities of each member of the organization should be specified in writing. If the company does not have an organization chart, this key recommends that one should be developed. For this reason, key number 2 is a key that comes before other improvement activities, and it is located outside on the lean circle (Fig. 9.1).

There are several methodologies for developing an organization chart, but, they are beyond the scope of this book. The basic idea is that the functions and responsibilities of each organization member have to be specified beforehand. Key number 2 recommends including some improvement requirements in each company function for planning and continuous process improvement.

The working conditions of employees also should be analyzed. Temporary workers and workers whose salaries are based on productivity (incentives) typically will be reluctant to cooperate in improvement actions because these actions are viewed as a threat that does not provide economic benefit for them and in many cases reduces the reward. In some companies, new employees sign a "moral contract" that forces them to improve their working conditions and productivity as an important part of their working requirements.

Improvement Team Activities

These improvements should be proposed and implemented by workers who are already familiar with the production area, in other words, teams that belong to the area in which the improvement project will be carried out. Without their cooperation, it will not be possible to

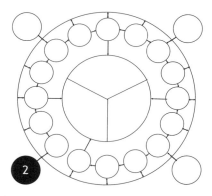

Figure 9.1. Location of the rationalizing the system tool in the 20 keys diagram.

make significant sustainable improvements. Frequently, company managers have neither the time nor the resources to maintain short-term improvements without the help of their workers.

Even in the consulting business, the cooperation of these working teams should be an early requirement. It is not unusual to hear factory employees comment, "Some time ago, a consultant came in for 5 months, fixing this problem, but we do not know how. Then the consultant left, and the area is back to where it was before the expert arrived."

Key number 3 (improvement team activities) is also outside the lean circle for the same reasons as the other outside keys. If the improvement team activities are not promoted, no permanent improvements will result (Fig. 9.2).

On the one hand, it is not the consultant's responsibility to create working teams. Team origination should come directly from factory managers. On the other hand, an experienced consultant will recognize the need for improvement teams to achieve long-term sustainable gains.

There is a very useful tool called *kaizen teian* that allows an organization to create improvement teams, but it is not the only method for developing teams. There are many other techniques to implement such programs; *kaizen teian* is only an example of such systems. *Teian* means "small suggestion." For this reason, *kaizen teian* can be translated as "small suggestions for continuous improvement." Besides the economic benefits that this system can provide, intangible improvements are also achieved, such as an increase in the morale of workers as well as their well-being.

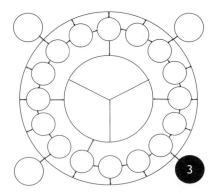

Figure 9.2. Location of the improvement team activities tool in the 20 keys diagram.

The team achievements should be rewarded in some way, but it does not always have to be an economic incentive. One observation from *kaizen teian* is that small rewards increase the number of suggestions. However, most suggestions are for small improvements that, in the long run, can outweigh big changes.

If a company decides to implement the *kaizen teian* system, it should have a suggestions revision system. It is important to avoid to use expressions such as "we are studying your idea" because workers easily can become discouraged. It is always necessary to provide a response to all suggestions, even if the new idea will not be considered for implementation.

Empowering Workers to Make Improvements

If the workers are required to propose and carry out improvements, it is necessary to teach them how to make those improvements. Key 14 oversees this strategy (Fig. 9.3).

Some workers gain tremendous insights about specific processes by using their machines over and over again. Some of these workers have been able to enhance the operation of these machines by extending the design complexity of some of the equipment to make them robotlike. Even without going this far, working on the same machines day after day offers the worker an opportunity to get to know the equipment more keenly and therefore to improve certain aspects of their operation. Key number 14 tries to globalize these local improvements so that they can be used in other areas.

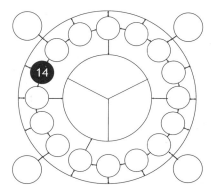

Figure 9.3. Location of the empowering workers to make improvements tool in the 20 keys diagram.

EFFICIENT MATERIALS USE–RELATED KEYS

Developing Your Suppliers

The relationship with suppliers cannot be limited to purchasing, delivery, inspection, and pricing commodities. Suppliers need to show a strong commitment to the supply due date, as well as work with the company to understand the functional requirements of their products so that the purchased products meet the needs of the company. The objective of the key number 12 is to facilitate the development of the products by suppliers. Frequently, small changes in the design of a component facilitate its manufacture and can reduce both cost and lead time (Fig. 9.4).

In addition, for its clients, it is important not to forget that the company is also a supplier. The relationship between the company and its supplier is the same the one that exists between the company and its customers.

Conserving Energy and Materials

Several small details taken together can represent big savings for a company. The right way to discover these savings consists of analyzing the aggregate costs. When the aggregate costs are analyzed, the most important costs can be eliminated. This is the goal of the key number 19 (Fig. 9.5).

The cost of materials procurement can be reduced in an important way by adding small savings that individually appear to have no importance or impact. Similarly, a single operation may not use energy

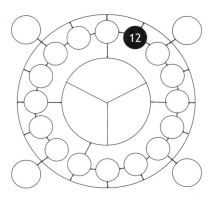

Figure 9.4. Location of the developing your suppliers tool in the 20 keys diagram.

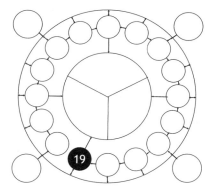

Figure 9.5. Location of the conserving energy and materials tool in the 20 keys diagram.

optimally yet still may appear to run efficiently. The repetitiveness of some operations can represent a significant energy waste. It is important to remember that sooner or later, insignificant events can add up and have a significant effect on a company.

Raw materials that are not used for production typically are sold at a lower price than they cost. If the company makes good use of its raw material, it can decrease both its raw materials costs and, as a result, its end-item cost (Fig. 9.6).

Reducing Inventory

After application of the preceding keys, inventory should decrease. For example, by improving operation through the more efficient use of raw materials, the quantity of raw materials required to produce the same number of articles decreases.

Figure 9.6. Example of a better use of raw materials.

In the case of on-hand stock usually held by the company, the inventory level will decrease proportionate to the reduction in raw materials use. Key number 4 addresses inventory-level reduction (Fig. 9.7).

It is important to remember that work-in-process is a type of inventory, and it depends in part on the factory layout. As a consequence, layout improvements can bring about reductions in inventory. Moreover, if the product lead time depends directly on the work-in-process, it also will decrease. As a consequence, it is possible to reduce delivery times.

VISUAL CONTROL–RELATED KEYS

Visual control makes it possible to identify and discover, by visual inspection, areas where there are problems and waste. This section analyzes some tools for visual control.

The goal of visual control is to manage only production and inventory anomalies in a manner similar to the way cowboys only control cattle that wander from the herd. Visual control, as a general concept, is an improvement tool included in the just-in-time (JIT) and in 20 keys (keys number 10 and 17) schemas (Fig. 9.8).

In this section, the *andon* and *kanban* visual control tools will be presented. In Chapter 8, a tool called the *sign strategy* was presented, and this is also a visual control tool. In fact, pictorial panels also can be used to show improvements achieved or to expose defective products (Fig. 9.9). The panels are also used to show the activity planning and

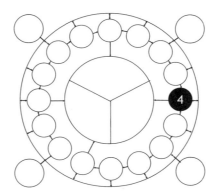

Figure 9.7. Location of the reducing inventory tool in the 20 keys diagram.

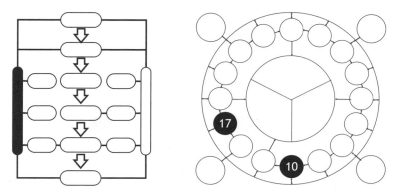

Figure 9.8. Visual control location in just-in-time and 20 keys diagrams.

the on-course processes. Frequently, panels are used to show weekly production rates.

Andon

One of the most popular visual control tools is a system called *andon,* which typically uses "alarm lights." These lights are used to indicate or warn workers of an activity that is going wrong.

Andon also can be used to detect material shortages (Fig. 9.10). The worker, by turning a light on, can let a supervisor know the trouble spot on the line that is causing the problem.

Kanban

The JIT philosophy changes the materials flow. Ohno defined Toyota's materials flow by looking at the way an American supermarket works.

Figure 9.9. Exhibiting defects is another visual control tool.

Figure 9.10. *Andon* is useful in detecting problems or material shortages in an assembly line.

Theoretically, when a component is used to manufacture or assemble one product, it generates a production or purchase order. This is called a *pull system* because the production orders flow from a finished good to raw materials, or control is *pulled* from assembly back to the basic processes. The other, or more conventional, system is called a *push system* (used in Occidental factories). In push systems, the production orders flow from raw materials to finished goods.

Pull systems created a material handling problem that Ohno had to solve. That is, how does the preceding process know what the following process wants? In order to answer this question, Ohno invented one of the best-known lean manufacturing tools, called *kanban*.

Sometimes people mistake *kanban* for JIT, or people think that *kanban* is the main lean manufactuing tool. However, the JIT and 20 keys models (Fig. 9.11) consider *kanban* to be just another tool, similar to the rest of the lean manufacturing tools.

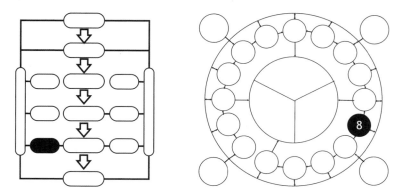

Figure 9.11. *Kanban* location in the JIT and 20 keys diagrams.

Kanban generally is translated as "card," and this is what *kanban* really is. It is a card located in a part's container providing information about the part, the preceding process, the quantity needed to fill the container, the container type, etc.

Another important mission of a *kanban* system is to establish the quantity of work-in-process. The number of *kanbans* is determined by management; workers cannot produce more containers than the number of *kanbans* issued.

Management by means of *kanban* has very strict rules that all workers should follow:

1. Do not to send defective products to the following process.
2. The trailing or following process removes the product from the current machine and leaves the *kanban*.
3. Produce only the quantity removed (the number of pieces written in the *kanban*).
4. Production must be level.
5. The *kanban* is used to stabilize the production process.

When demand is not level, it is not possible to use the *kanban* system because the number of *kanbans* needed would change constantly. In such cases, it is necessary to use other flow-control methodologies, such as the traditional stock management techniques, based on safety stock and basic economic order quantity (EOQ).

Theoretical Operation of the Kanban System. The theoretical operation of a *kanban* system is very simple. Suppose that a system is in steady state, where no one manufactures a product until one finished good is sold. Although our example is for a greatly simplified manufacturing process for, say, a car, the example will explain, in a simple way, the operation of the *kanban* system (Fig. 9.12).

Figure 9.12. Starting situation in a *kanban* operation cycle.

When a car is sold, the car's *kanban* moves immediately to the previous workstation (Fig. 9.13). To finish the car assembly, it is necessary to have a chassis painted the color required by the buyer, four wheels, and one painted roof. When the chassis is removed, its *kanban* is sent to the chassis section (Fig. 9.14).

Once the wheels and the roof are assembled on the car, a new vehicle will be ready for sale (Fig. 9.15). In this example, the wheels container had eight wheels, so the *kanban* is not removed until the four remaining wheels are removed.

When a new chassis arrives from the preceding section, the process stops again (Fig. 9.16). The next car sold will repeat this *kanban* cycle. This process will be repeated in all the workstations from finished goods to raw materials, generating the production or purchasing orders.

Kanban in Practice. The production flexibility demanded by the JIT philosophy limits the use of a pure *kanban* system. In practice, a *kanban* system is used for the self-management of manufacturing cells. That is to say, it is used for the autonomous scheduling of workstations. Each cell could manufacture different components if the worker chooses what product to manufacture next according to the *kanban* board (Fig. 9.17).

The board contains the *kanbans* provided from preceding sections. *Kanbans* are accumulated in each column based on the consumption rate. Each column on the board (relates to one product reference) is divided in three regions:

- *Below the green line.* If *kanbans* are below the first level, then it is not necessary to manufacture the product (especially if there are other priority items that the worker must manufacture).
- *Between green and red lines.* When *kanbans* are located between the two colored lines, the worker should schedule the product to fill the containers and replace the stock in the next section.

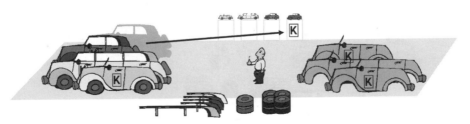

Figure 9.13. A car sale starts the *kanban* operation cycle.

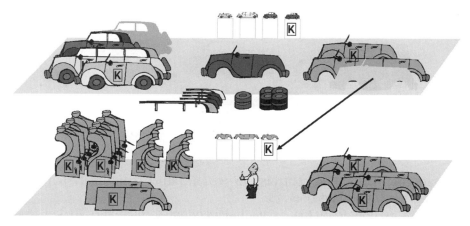

Figure 9.14. The kanban system moves *kanbans* through sections.

- *Over the red line.* If the number of *kanbans* exceeds the red line, the next assembly cell will consume the safety-stock containers. In this case, the worker should stop manufacturing the current product and start manufacturing the product that is below the red line.

This is a method for transforming the scheduling problem into a visual control methodology. This method is quite similar to the traditional system of warehouse management based on order point and safety stock. However, in this case, the worker is able to decide the scheduling sequence. Finally, a *kanban* does not have to be a card. Sometimes any object or signal is enough to show the need for a certain article (Fig. 9.18).

TECHNOLOGY-RELATED KEYS

Jidoka

Jidoka is translated as "autonomation." It is a word that consists of two concepts. On the one hand is the autonomy of control of the equipment (literal translation of *jidoka*). On the other hand is automation of processes in a simple way.

Since few unmanned production systems exist, *jidoka* has taken on an additional character of integrating the human into an automated system so that the human is well integrated into the system. It is im-

Figure 9.15. After the assembly operation, the new car is finished.

portant that the human controls the system rather than the system controlling the individual operator.

Jidoka is contemplated as an independent tool in the 20 keys, as well as in the JIT philosophy. In the lean manufacturing schema, *jidoka* is one of the three main pillars (Fig. 9.19).

Equipment improvements, by means of automation, require capital investment. Unless effective working procedures have been improved already before automation is brought on board, the capital investment can be larger than necessary, and it can result in automating an *inefficient* activity (waste).

The main goal of machine autonomy control is having machines that manufacture without producing waste (Fig. 9.20), including any defects. According to the *jidoka* tool, it is necessary to make the machines think (*ninben no tsuita jidoka*).

Jidoka is not a utopia. In most manufacturing and even domestic environments, there are numerous cases of autonomation. For instance, who looks at the washing machine window while it is washing the clothes or watches the oven while a chicken is roasted?

Ohno reached the point where he painted the computer numerical control (CNC) windows to avoid workers looking at how the machines were working. *Jidoka's* technique looks for a machine automation as

Figure 9.16. The factory is once again in a steady state.

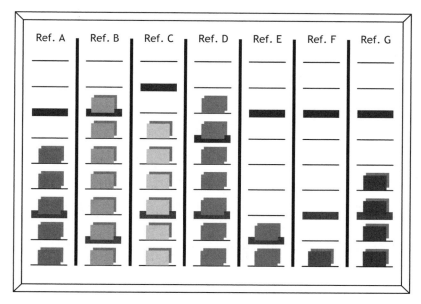

Figure 9.17. Kanban board.

simple as possible, i.e., machines that can be started or stopped with a single button (Fig. 9.21).

Using Information Systems

The use of microprocessors should not be limited only to computers and programs that help to manage the factory. Key number 18 presents

Figure 9.18. Kanban also can be an object such as a cloth or other visual signal.

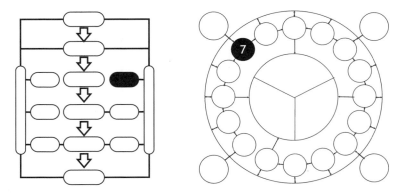

Figure 9.19. Location of the *jidoka* in the JIT and 20 keys diagrams.

several examples of information systems and electronic device utilization (Fig. 9.22).

Electronic devices can be found in many production applications, such as sensors, robots, data-detection systems, etc. *Jidoka* is a particular case of automation and even in the 20 keys structure is addressed independently.

Leading Technology and Site Technology

Leading technology is the group of skills that the improvement team uses during development of the manufacturing process and its improvements. It is also called *know-how*. Key number 20, the last key, emphasizes the importance of managing the company using the *know-how philosophy* (Fig. 9.23).

Figure 9.20. Automation means making machines think.

Figure 9.21. *Jidoka* bets for the simplest automation. The easier, the better.

The systematic application of the 20 improvement keys forces a company to introduce a continuous improvement process. Kobayashi concludes his book by saying that thanks to this continuous improvement process, the company is prepared to last for generations, becoming a sector leader.

SUMMARY

The objective of this chapter was to present other tools mentioned in Kobayashi's book that had not been explained in previous chapters. These improvement methodologies have been grouped into four characteristics: human resources–related keys, efficient materials use–related keys, visual control–related keys, and technology-related keys.

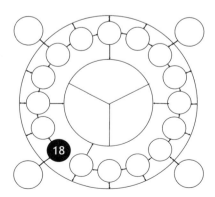

Figure 9.22. Location of the using information systems tool in the 20 keys diagram.

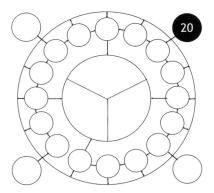

Figure 9.23. Location of the leading technology and site technology tool in the 20 keys diagram.

RECOMMENDED READINGS

Japan Human Relatins Association, *Kaizen Teian 1: Developing Systems for Continuous Improvement through Employee Suggestions.* Portland, OR: Productivity Press, 1992.

Japan Management Association, *Kanban Just-in-Time at Toyota: Management Begins at the Workplace.* Portland, OR: Productivity Press, 1989.

Appendix A:

Problems

INTRODUCTION

Numeric problems are classified in five different categories based on the chapers of this book. Some chapters don't have specific numeric problems, but some concepts can appear in the problems of other chapters.

Some problems may seem to be the same, but it is important to read the data carefully to notice small changes. The objective of these small changes is to prove how a simple change in data can lead to a different solution. Translating this idea to real-world problems will show that it is very difficult to implement standard solutions to every type of problem.

At the end of this appendix, the solutions are included. Also included are some reference problems that show how to solve the problems. Some of the problems presented herein do not have a unique solution. In other words, there are several possible ways to solve such problems. Therefore, if your answer does no coincidence with the one provided, this does not mean that you have not solved the problem properly. Some of the answers include such sentences as, "Look at the theory" or "Graphic answer," which means that the solution is not included.

CONTINUOUS IMPROVEMENT TOOLS

Problem IT1

Breakdowns and defective products that were manufactured for the last 15 days on a particular machining center, the CNC-27 machine, are presented in the following table:

Day	1	2	3	4	5	6	7	8	9	10	11	12	13	14	15
Breakdown code	A	—	B	B	C	D	D	—	—	—	E	—	—	F	G
Breakdown start time	3	—	16	—	10	17	—	—	—	—	6	—	—	15	3
Breakdown end time	5	—	—	2	13	—	3	—	—	—	12	—	—	21	8
Defective products	180	320	25	150	100	380	210	0	5	35	100	32	23	100	130

The breakdown code represents each recorded breakdown. If the same code appears on more than one day, it is because the repair did not take place or was not finished on the preceding day.

The machine's total setup time is 187 minutes, and in the 15 days that were investigated the machine produced 6280 correct products. The company works 24/7. The theoretical cycle time is 2 minutes per part.

1. What is the overall equipment efficiency (OEE) value?
2. What are the quality, performance, and availability rates?
3. Which measure would you apply first to improve the OEE indicator? Justify your answer.

Problem IT2

Breakdowns and defective products manufactured over the last 15 days on a CNC-27 machine are presented in the following table:

Day	1	2	3	4	5	6	7	8	9	10	11	12	13	14	15
Breakdown code	A	—	B	B	C	D	D	—	—	—	E	—	—	F	G
Breakdown start time	3	—	16	—	10	17	—	—	—	—	6	—	—	15	3
Breakdown end time	5	—	—	2	13	—	3	—	—	—	12	—	—	21	8
Defective products	180	320	25	150	100	380	210	0	5	35	100	32	23	100	130

The breakdown code represents each recorded breakdown. If the same code appears on more than one day, it is because the repair did not take place or was not finished on the preceding day.

The machine's total setup time is 187 minutes, and in those 15 days the machine produced a total of 6280 products. The company works 24/7. The theoretical cycle time is 2 minutes per part.

1. What is the overall equipment efficiency (OEE) value?
2. What are the quality, performance, and availability rates?
3. Which measure would you apply first to improve the indicator OEE? Justify your answer briefly.

Problem IT3

(This problem is recommended after Chap. 6.) Breakdowns and defective products manufactured over the last 15 days on a CNC-27 machine are presented in the following table:

Day	1	2	3	4	5	6	7	8	9	10	11	12	13	14	15
Breakdown code	A	—	B	B	C	D	D	—	—	—	E	—	—	F	G
Breakdown start time	3	—	16	—	10	17	—	—	—	—	6	—	—	15	3
Breakdown end time	5	—	—	2	13	—	3	—	—	—	12	—	—	21	8
Defective products	180	320	25	150	100	380	210	0	5	35	100	32	23	100	130

The breakdown code represents each recorded breakdown. If the same code appears on more than one day, it is because the repair did not take place or was not finished on the preceding day. On the eighth day, a 24-hour scheduled stop was carried out for preventive maintenance tasks.

The machine's total setup time is 187 minutes, and in those 15 days the machine produced a total of 6280 products. The company works 24/7. The theoretical cycle time is 2 minutes per part.

1. What is the overall equipment efficiency (OEE) value?
2. What are the quality, performance, and availability rates?
3. Which measure would you apply first to improve the OEE indicator? Justify your answer briefly.

Problem IT4

(This problem is recommended after Chap. 7.) In a production cell dedicated to the manufacture of a specific product family, the following data corresponding to 10 workdays were obtained. The cell works 8 hours a day.

In the 10 days that were studied, 6 hours was planned and used for preventive maintenance. Breakdowns consumed 4 hours. The total production was 20,280 products, but 400 of those products were defects. The theoretical daily production is 2800 products.

After a SMED implementation, the following data were collected for a similar 10-day period. In those 10 days, 6 hours of preventive maintenance also was used. Breakdowns consumed 5 hours. The total production was 20,100 units, including 100 defect products. The theoretical daily production continues to be 2800 products.

1. How has the SMED implementation varied the overall equipment efficiency (OEE) rate?
2. Is the result obtained consistent with the SMED implementation? Why?
3. Why has the number of defects decreased?

Problem IT5

In a production cell dedicated to the production of a product family, the following data corresponding to 10 working days have been obtained (the working time is 24/7). The theoretical production was 14,500 units per day.

Day	1	2	3	4	5	6	7	8	9	10
Breakdowns (h)	—	2	5	—	—	—	2	—	3	—
Setup time (h)	2	1	—	2	2	—	2	3	4	—
Defects (units)	100	150	200	100	150	220	130	80	140	200
Total production (units)	12,000	13,280	11,300	12,100	14,000	13,000	12,300	11,900	13,100	11,700

After a large technical investment, a new theoretical production rate of 20,000 units per day was fixed. After 10 working days, the following data were collected:

Day	1	2	3	4	5	6	7	8	9	10
Breakdowns (h)	—	3	—	—	4	—	4	—	1	—
Setup time (h)	—	3	—	2	—	2	3	2	2	2
Defects (units)	200	250	100	130	230	125	240	320	100	180
Total production (units)	15,000	13,200	15,600	14,800	15,200	14,800	12,000	15,600	15,100	14,900

1. How has this changed the overall equipment efficiency (OEE) rate?
2. Has this change been worth the technical investment?

Problem IT6

(This problem is recommended after Chap. 6.) In a production cell dedicated to the production of a product family, the following data corresponding to 10 working days have been obtained (the working time is 24/7). The theoretical production was 90 units per day.

Day	1	2	3	4	5	6	7	8	9	10
Breakdowns (h)	3	1	2	—	1	1	2	—	5	—
Setup time (h)	—	4	—	—	—	4	—	—	—	4
Defects (units)	6	7	5	3	5	3	3	6	4	3
Total production (units)	88	70	78	83	80	60	83	77	75	79

After a maintenance improvement project, it was decided to make a daily planned stop of 2 hours to carry out preventive maintenance tasks. After 10 working days, the following data were collected:

Day	1	2	3	4	5	6	7	8	9	10
Breakdowns (h)	1	—	—	—	2	—	—	—	3	—
Setup time (h)	—	4	—	—	—	4	—	—	—	4
Defects (units)	5	8	3	4	3	5	3	2	2	7
Total production (units)	77	72	77	80	70	75	80	72	77	75

1. How has this changed the overall equipment efficiency (OEE) rate?
2. Was the project worthwhile? Why?

Problem IT7

(This problem is recommended after Chap. 4.) In a production cell dedicated to the production of a product family, the following data corresponding to 10 working days have been obtained (the working time is 24/7). The theoretical production was 90 units per day.

Day	1	2	3	4	5	6	7	8	9	10
Breakdowns (h)	—	2	—	—	5	—	3	—	—	6
Setup time (h)	—	—	3	—	—	3	—	—	3	—
Defects (units)	7	6	8	2	1	9	6	5	5	6
Total production (units)	70	65	85	80	63	70	73	68	72	67

In order to avoid defects, a *poka-yoke* device and a source inspection project were carried out. The next day, after implementation of those strategies, data from 10 working days were collected:

Day	1	2	3	4	5	6	7	8	9	10
Breakdowns (h)	3	—	—	4	—	1	—	7	—	1
Setup time (h)	—	—	3	—	—	3	—	—	3	—
Defects (units)	16	14	11	11	8	8	5	5	3	2
Total production (units)	81	70	65	67	72	80	83	71	67	69

1. How has this changed the overall equipment efficiency (OEE) rate?
2. Were the strategies worthwhile? Why?

Problem IT8

Breakdowns and defective products manufactured over the last 15 days on a CNC-27 machine are presented in the following table:

Day	1	2	3	4	5	6	7	8	9	10	11	12	13	14	15
Breakdowns (h)	A	—	B	B	C	D	D	—	—	—	E	—	—	F	G
Setup time (h)	3	—	16	—	10	17	—	—	—	—	6	—	—	15	3
Defects (units)	5	—	—	2	13	—	3	—	—	—	12	—	—	21	8
Total production (units)	180	320	25	150	100	380	210	0	5	35	100	32	23	100	130

The breakdown code represents each recorded breakdown. If the same code appears on more than one day, it is because the repair did not take place or was not finished on the preceding day.

The machine's total setup time is 187 minutes, and in the 15 days investigated, the machine produced 6280 correct products. In addition, 360 defective products were recovered after being reprocessed.

The company works 24/7. The theoretical cycle time is 2 minutes per part.

1. What is the overall equipment efficiency (OEE) value?
2. What are the quality, performance, and availability rates?
3. Which measure would you apply first to improve the OEE indicator? Justify your answer.

FACILITIES LAYOUT

Problem FL1

A company is dedicated to manufacturing parts 24 hours a day. Data corresponding to products, processes, and equipment are shown in the following tables. Another table with a historical order summary is also presented. These orders correspond to a period of time considered as representative for normal production.

Sections			Bill of Operations		
Code	Name	Number of Machines	Product	Operation (Sequential Order)	Section Code
1	Milling	1	A	1	1
2	Lathes	2	A	2	2
3	Final control	1	A	3	3
			B	1	2

Historical Summary				
		B	2	1
		B	3	3
Product	Orders			
A	50	C	1	1
B	30	C	2	3
C	10	D	1	2
D	10	D	2	3

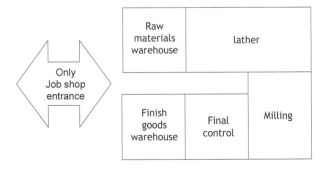

Current layout (drawing is in scale dimension)

1. Create the transfer matrix for the products and machines shown.
2. Decide if it is necessary to modify the current layout. If so, draw the new layout.

Problem FL2

Precision Corp. is a company dedicated to parts manufacturing. Its facility layout is shown in the following figure. Some important data also are shown in the following tables.

The company managers want to carry out a methods and time study to reduce the idle time. As a previous step, the managers needed to analyze the process fowchart of the most important parts.

Sections				Bill of Operations of Part P7			
Section	Machine Codes	Machine Type	Part	Operation (Sequential Order)	Self-Control	Section Code
1	M11, M12	Milling	P7	1	No	2
2	M21	Lathe	P7	2	Yes	1
3	M31, M32	Grinding	P7	3	Yes	2
4	—	Quality	P7	4	No	3
			P7	5	—	4

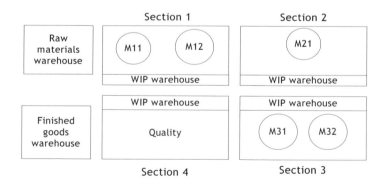

1. Indicate what layout type the company has. Justify your answer, and indicate the advantages and disadvantages of this layout type.
2. Draw the process flowchart for part P7.

Problem FL3

The following tables show data corresponding to the production process of five products.

Product	Operation (Sequential Order)	Section Code	Product	Operation (Sequential Order)	Section Code
A	1	1	D	1	2
A	2	2	D	2	3
A	3	1	D	3	2
A	4	3	D	4	4
B	1	1	E	1	2
B	2	2	E	2	3
B	3	4	E	3	4
C	1	2			
C	2	4			

Create the transfer matrix.

Problem FL4

The managers of a company want to carry out a methods and time study to reduce the idle time. As a previous step, the managers needed to analyze the process flowchart for the most important products. The process and the facility layout are shown in the following figure. The following tables show data corresponding to the four products.

Product	Operation (Sequential Order)	Section Code	Product	Operation (Sequential Order)	Section Code
A	1	1	D	1	2
A	2	2	D	2	3
A	3	1	D	3	2
A	4	3	D	4	4
B	1	1			
B	2	2			
B	3	4			

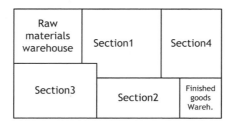

Draw the process flowchart for the products.

CELLULAR LAYOUT

Problem CL1

In the following table, some of the time-study data for a production facility are presented. This time study is being carried out to establish the standard time of an operation using the continuous-timing method. For this study, each day the worker has 420 minutes of a total of 480 minutes to carry out production activities.

Task		1	2	3	4	5	6	7	8	9	10	AF
1	E											0.95
	A	11	45	81	113	148	183	220	255	293	330	
2	E											1.1
	A	28	63	97	130	165	200	236	275	310	348	
3	E											1.05
	A	35	70	104	137	173	208	244	282	318	355	

Determine the standard time for the task and the number of products manufactured each hour.

Problem CL2

A production line with 12 tasks has the precedence relationships and durations specified of the following table. The production line works 8 hours a day, and the daily demand is 40 products.

Task	Precedence	Duration (min)
A	—	6
B	—	9
C	A	4
D	A	5
E	B	4
F	C	2
G	C, D	3
H	F	7
I	G	3
J	E, I	1
K	H, J	10
L	K	1

1. Assign the tasks according to the total-number-of-following-tasks heuristic.

2. Assign the tasks according to the individual-durations heuristic.
3. Assign the tasks according to the largest-positional-weight heuristic.

Problem CL3

A product will be manufactured in an U-shaped cell. According to a market study, the product daily demand will be approximately 480 units. The cell will operate 8 hours/day. The tasks that should be carried out are shown in the following table:

Task	Duration (s)	Precedence
A	16	—
B	18	A
C	13	B
D	14	—
E	10	D
F	12	C
G	7	C
H	11	E
I	14	E
J	17	F, G, H, I
K	11	J
L	20	K

1. Draw the line balancing in order to minimize the number of workstations.
2. Determine the total idle time, the lead time, and the work-in-process.
3. Make an outline of the cell layout indicating what tasks will be carried out in each station, and make a Gantt chart of the cell's operation.

Problem CL4

A product will be manufactured in a U-shaped cell. According to a market study, the product's daily demand will be approximately 105 units. The cell will run 24 hours/day, but it is expected that 12.5 percent of time the cell will be stopped owing to maintenance, defects, or other problems. The tasks that should be carried out are shown in the following table:

Task	Duration (min)	Precedence
A	6	—
B	2	—
C	4	A
D	5	A
E	4	B
F	2	C
G	3	C, D
H	3	F
I	5	G
J	1	E, I
K	8	H, J
L	4	K

1. Draw the line balancing in order to minimize the number of work-stations and work-in-progress according to the total-number-of-following-tasks heuristic.
2. Determine the total idle time, the lead time, and the work-in-process.
3. Make a sketch of the cell layout indicating what tasks will be carried out in each station by the assigned workers, and then develop a cell Gantt diagram.

Problem CL5

A product will be manufactured on a production line. According to a market study, the daily demand for the product will be 48 units. The line will run 8 hours/day. The tasks to be carried out are shown in the following table:

Task	Duration (min)	Precedence
A	10	—
B	7	A
C	8	A
D	4	C
E	7	C
F	3	B
G	16	F, D, E
H	2	G
I	1	G
J	6	H, I

1. Draw the line balancing in order to minimize the number of work-stations according to the individual-durations heuristic.

2. Determine the total idle time, the lead time, and the work-in-process.
3. Make an outline of the cell layout indicating what tasks will be carried out in each station by the assigned workers.

Problem CL6

A product will be manufactured on a production line. The line *takt* time is 18 seconds. The tasks that should be carried out are shown in the following table:

Task	Duration (s)	Precedence
A	5	—
B	4	A, E
C	2	—
D	10	E
E	5	—
F	3	C, D
G	5	H, I
H	7	—
I	8	—
J	4	B, F, G

1. Draw the line balance in order to minimize the number of workstations according to the largest-positional-weight heuristic.
2. Determine the total idle time, the lead time, and the work-in-process.
3. Make a sketch of the cell layout indicating what tasks will be carried out in each station by the assigned workers.

Problem CL7

A product will be manufactured on a production line. The daily demand for the product will be 180 units. The line will run 24 hours/day. The tasks that should be carried out are shown in the following table:

Task	Duration (min)	Precedence
A	5	—
B	5	—
C	4	A, B
D	1	C
E	1	C
F	1	D
G	2	D, E
H	1	G
I	1	H

1. Draw the line balance in order to minimize the number of work-stations.
2. Determine the total idle time, the lead time, and the work-in-process.
3. Make a sketch of the cell layout indicating what tasks will be carried out in each station.

Problem CL8

A product will be manufactured in a U-shaped cell. The product daily demand will be 480 units. All operations are considered assembly tasks, that is to say, they do not need any machines. The cell will run 8 hours/day. The tasks that should be carried out are shown in the followng table:

Task	Duration (s)	Precedence
A	18	—
B	16	A
C	18	B
D	13	C
E	14	A
F	10	E
G	12	D
H	7	D
I	11	F
J	14	F
K	17	G, H, I, J
L	11	K
M	10	L

1. Draw the line balance in order to minimize the number of work-stations according to the individual-durations heuristic.
2. Determine the total idle time, the lead time, and the work-in-process.
3. Make a sketch of the cell layout indicating what tasks will be carried out in each station by the assigned workers, and make the cell Gantt diagram.

Problem CL9

One company wants to design a U-shaped cell to manufacture a product that will have a demand of 42 units/day. The line will work 7 hours/

day. The tasks that should be carried out are shown in the following table:

Task	Duration (min)	Precedence
A	5	—
B	1	A, C
C	1	—
D	3	C
E	12	B, D
F	3	B, E
G	1	D, F
H	1	G
I	5	G
J	3	H, I

1. Draw the line balance in order to minimize the number of work-stations according to the largest-positional-weight heuristic.
2. Determine the total idle time, the lead time, and the work-in-process.
3. Make a sketch of the cell layout indicating what tasks will be carried out in each station by the assigned workers.
4. What advantages does this U-shaped layout have? Would the U-shaped layout be recommended to manufacture this product?

Problem CL10

A company that manufactures combo-TVs (TV + DVD) wants to have two assembly cells owing to component space restrictions. In one cell, the TV will be assembled, and in the other cell, the DVD will be assembled. It is also necessary to have another task to combine the two parts to get the final product. The daily demand is 12 units, and both lines will run 8 hours a day. The tasks that should be carried out are shown in the following table:

TV Assembly			DVD Assembly			Final Assembly		
Task	Precedence	t_i (min)	Task	Precedence	t_i (min)	Task	Precedence	t_i (min)
T1	—	20	V1	—	10	MF1	T6, V7	15
T2	T1, T3	15	V2	V1	5			
T3	—	25	V3	V1, V2	15			
T4	T2	15	V4	V2, V3	23			
T5	T4, T3	10	V5	V4	20			
T6	T4, T5	5	V6	V5, V3	10			
			V7	V6	18			

1. Draw the line balance in order to minimize the number of work-stations according to the total-number-of-following-tasks heuristic.
2. Determine the total idle time.
3. Make a sketch of the cell layout indicating what tasks will be carried out in each station by the assigned workers, and draw a Gantt chart for the cell.

Problem CL11

A product will be manufactured in a U-shaped cell. The cell has to manufacture one product every 40 minutes during an 8-hour shift every day. All the operations are assembly tasks; that is to say, they do not need any machines. The following table shows the tasks:

Task	Duration (min)	Precedence
A	12	—
B	6	A, C
C	10	—
D	5	C
E	6	B, D
F	17	E
G	8	D
H	4	E, G
I	13 (automatic)	F, H
J	2	I
K	3	J
L	15 (automatic)	K
M	1	L

Tasks I and L are carried out automatically, i.e., without human intervention. Tasks H, J, K, and M are part setting and removing tasks, where human intervention is needed (tasks H and J of the I operation; tasks K and M of the L operation).

1. Draw the line balance in order to minimize the number of work-stations according to the individual-durations heuristic.
2. Determine the minimum number of workers assigned to the cell.
3. Determine the total idle time, the lead time, and the work-in-process.

4. Make a sketch of the cell layout indicating what tasks will be carried out in each station by the assigned workers, and make a cell Gantt diagram.

Problem CL12

A product will be manufactured in a U-shaped cell. The product daily demand will be 480 units. The operations are all assembly tasks; that is to say, they do not need any machines. The cell will run 8 hours/day. The tasks that should be carried out are shown in the following table:

Task	Duration (s)	Precedence
A	18	—
B	16	A
C	75	B
D	13	C
E	14	A
F	10	E
G	12	D
H	7	D
I	11	F
J	14	F
K	17	G, H, I, J
L	11	K
M	10	L

1. Draw the line balance in order to minimize the number of workstations according to the individual-durations heuristic.
2. Determine the total idle time, the lead time, and the work-in-process.
3. Make a sketch of the cell layout indicating what tasks will be carried out in each station by the assigned workers, and make a Gantt chart for the cell.

Problem CL13

A product will be manufactured in a U-shaped cell. The product daily demand will be 1680 units. The cell will work two shifts of 8 hours each, but it is expected that 12.5 percent of time the cell will be stopped owing to maintenance. The tasks that should be carried out are shown in the following table:

Task	Duration (s)	Precedence
A	8	—
B	12	A
C	6	—
D	10	B, C
E	5	D, A
F	40	D
G	14	E, F
H	5	F
I	6	H, G
J	10	I
K	9	I
L	5	H, J
M	14	K, L

1. Draw the line balance in order to minimize the number of work-stations according to the individual-durations heuristic.
2. Determine the total idle time, the lead time, and the work-in-process.
3. Make a sketch of the cell layout indicating what tasks will be carried out in each station by assigned workers, and develop a Gantt diagram of the cell.

Problem CL14

A product will be manufactured in a U-shaped cell. All the operations are assembly tasks; that is to say, they do not need any machines. The line will produce 3 products per hour. The following table shows the data:

Task	Duration (min)	Precedence
A	16	—
B	5	A
C	5	A
D	8	A
E	3	C, D
F	6	D
G	1	B, F
H	3	D, E
I	10	G, H

1. Draw the line balance in order to minimize the number of work-stations according to the total-number-of-following-tasks heuristic. Also, make a cell Gantt diagram to justify your answer.
2. Determine the total idle time, the lead time, and the work-in-process.

3. Make a sketch of the cell layout indicating what tasks will be carried out in each station by the assigned workers.

Problem CL15

A production line with 12 tasks has the precedence relationships and durations presented in the following table. The line runs 8 hours a day and has a daily demand of 80 products, but a demand reduction is anticipated that will reduce demand to half its current level.

Task	Precedence	Duration (min)
A	—	6
B	—	9
C	A	4
D	A	5
E	B	4
F	C	2
G	C, D	3
H	F	7
I	G	3
J	E, I	1
K	H, J	10
L	K	1

1. Draw the line balance in order to minimize the number of work-stations according to the largest-positional-weight heuristic, and make a Gantt diagram for the cell.
2. Determine the total idle time, the lead time, and the work-in-process.
3. Make a sketch of the cell layout indicating what tasks will be carried out in each station by the assigned workers.

Problem CL16

A production line with 12 tasks has the precedence relationships and durations shown in the following table:

Task	Precedence	Duration (min)
A	—	6
B	—	9
C	A	4
D	A	5
E	B	4

Task	Precedence	Duration (min)
F	C	2
G	C, D	3
H	F	7
I	G	3
J	E, I	1
K	H, J	10
L	K	1

The line works 8 hours a day, and the daily demand is 48 units. The company wants to use a U-shaped layout because its main objective is to get production flexibility.

1. Draw the line balancing according to the individual-durations heuristic so that the pursued objective is achieved.
2. Determine the work-in-process.
3. Make a skecth of the cell layout indicating what tasks will be carried out in each station by the assigned workers, and make a Gantt diagram for the cell.

Problem CL17

A product will be manufactured in a U-shaped cell. According to a market study, the product's daily demand will be approximately 24 units. The cell will work 8 hours/day. The tasks that should be carried out are shown in the following table:

Task	Duration (min)	Precedence
A	10	—
B	7	A
C	8	A
D	4	C
E	7	C
F	3	B
G	16	F, D, E
H	2	G
I	1	G
J	6	H, I

1. Draw the line balance in order to minimize the number of work-stations according to the individual-durations heuristic.

2. Determine the total idle time, the lead time, and the work-in-process.
3. Make a sketch of the cell layout indicating what tasks will be carried out in each station by the assigned workers, and make a Gantt diagram for the cell.

Problem CL18

A product will be manufactured in a U-shaped cell. According to a market study, the product daily demand will be approximately 24 units. The cell will work 8 hours/day. The tasks that should be carried out are shown in the following table:

Task	Duration (min)	Precedence
A	10	—
B	7	A
C	8	A
D	4	C
E	7	C
F	3	B
G	16	F, D, E
H	2	G
I	1	G
J	6	H, I

1. Draw the line balance in order to minimize the number of workstations according to the largest-positional-weight heuristic.
2. Determine the total idle time, the lead time, and the work-in-process.
3. Make a sketch of the cell layout indicating what tasks will be carried out in each station by the assigned workers, and make a Gantt diagram for the cell.

Problem CL19

A product will be manufactured in a U-shaped cell, and the product daily demand will be 24 units. The cell will run 8 hours/day. The tasks that should be carried out are shown in the following table:

Task	Duration (min)	Precedence
A	10	—
B	8	A, C
C	3	—
D	1	B
E	7	B
F	2	B
G	16	F, D, E
H	6	G
I	2	H
J	1	H

1. Draw the line balance in order to minimize the number of work-stations according to the individual-durations heuristic.
2. Determine the total idle time, the lead time, and the work-in-process.
3. Make a sketch of the cell layout indicating what tasks will be carried out in each station by the assigned workers, and make a Gantt diagram for the cell.

Problem CL20

A product will be manufactured in a U-shaped cell, and the product hourly demand will be 3 units. The cell will work 8 hours/day. The tasks that should be carried out are shown in the following table:

Task	Duration (min)	Precedence
A	7	—
B	1	A, D
C	7	D
D	4	—
E	4	B, C
F	1	C
G	6	F, H
H	7	E
I	3	H
J	5	G
K	5	G, I
L	5	I

1. Draw the line balance in order to minimize the number of work-stations according to the individual-durations heuristic.

2. Determine the total idle time, the lead time, and the work-in-process.
3. Make a sketch of the cell layout indicating what tasks will be carried out in each station by the assigned workers, and make a Gantt diagram of the cell.

Problem CL21

A product will be manufactured in a U-shaped cell. According to a market study, the cell must produce one product each 14 minutes. The tasks that should be carried out are shown in the following table:

Task	Duration (min)	Precedence
A	11	—
B	4	A
C	4	A
D	5	A
E	3	B
F	1	B
G	4	C, D
H	2	E, F
I	9	F, G
J	6	H, I
K	1	J

1. Draw the line balance in order to minimize the number of work-stations according to the total-number-of-following-tasks heuristic.
2. Determine the total idle time.
3. Make a sketch of the cell layout indicating what tasks will be carried out in each station by the assigned workers, and make a Gantt chart for the cell.

Problem CL22

A product will be manufactured in a cell, and the product hourly demand will be 4 units. The tasks that should be carried out are shown in the following table:

Task	Duration (min)	Precedence
A	6	—
B	3	A
C	7	A
D	5	—
E	11	D
F	4	B, C, E
G	3	F
H	5	F
I	6	F
J	8	H, I
K	17	G, I
L	5	K, J

1. Draw the line balance in order to minimize the number of work-stations according to the largest-positional-weight heuristic.
2. Determine the total idle time.
3. Make a sketch of the cell layout indicating what tasks will be carried out in each station by the assigned workers, and make a Gantt diagram of the cell.

Problem CL23

A product will be manufactured in a U-shaped cell. According to a market study, the product daily demand will be approximately 96 units. The cell will work 16 hours/day. The tasks that should be carried out are shown in the following table:

Task	Duration (min)	Precedence
A	4	—
B	2	—
C	1	A
D	6	B
E	1	C
F	2	C
G	5	D, F
H	5	E, G
I	3	H
J	1	H

1. Draw the line balance in order to minimize the number of work-stations according to the individual-durations heuristic.

2. Determine the total idle time, the lead time, and the work-in-process.
3. Make a sketch of the cell layout indicating what tasks will be carried out in each station by the assigned workers, and make the Gantt diagram of the cell.

MAINTENANCE

Problem MN1

An engineering department is designing a new machine, and the engineers have to choose a component for a critical function; that is to say, if it breaks, the machine stops. The working team has to decide between two suppliers of similar components. The components present the following bathtub curves:

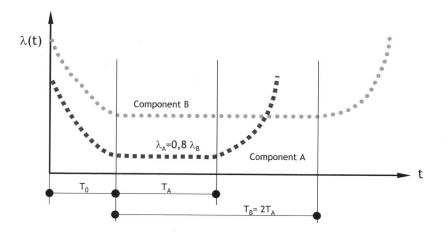

1. Which component or supplier should be chosen?
2. Elaborate a maintenance plan for the component.

Problem MN2

Breakdowns and defective products produced for the last 15 days on a CNC machine are presented in the following table:

Day	1	2	3	4	5	6	7	8	9	10	11	12	13	14	15
Breakdown code	A	—	B	B	C	D	D	—	—	—	E	—	—	F	G
Breakdown start time	3:00	—	16:00	—	10:00	17:00	—	—	—	—	6:00	—	—	15:00	3:00
Breakdown end time	5:00	—	—	2:00	11:00	—	3:00	—	—	—	12:00	—	—	21:00	8:00

The breakdown code represents each recorded breakdown. If the same code appears on more than one day, it is because the repair did not take place or was not finished on the preceding day. The factory works 24/7.

1. Calculate the machine statistical availability beginning at 0:00 hour on day 1 and finishing at 0:00 hour on day 16 (or 24:00 on day 15).
2. What is the difference between statistical availability and the overall equipment efficiency (OEE) rate?

Problem MN3

An engineering department is designing a new machine, and the engineers have to choose a component for an aesthetic function, that is to say, a function that is not critical for the machine. The working team has to decide between two suppliers of similar components. The components present the following bathtub curves:

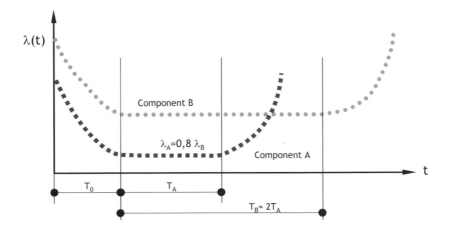

1. Which component or supplier should be chosen?
2. Elaborate a maintenance plan for the component.

MOTION STUDY

Problem MS1

A person has an electric toaster that can toast two slices of bread simultaneously, but only for one of the two slice faces (see the following drawing).

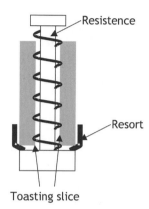

Since each door stays closed with a spring, it is necessary to hold it open while the slice is introduced. As a consequence, the person is only able to put in one slice at a time.

The person wants to toast three slices of bread in the shortest time possible. The operations and durations to carry out the process are the following:

Put in a slice	10 s
Toast a slice side	25 s
Turn the slice	10 s
Remove the slice	10 s

1. How much time is required to toast the three slices on both faces?
2. Show your solution with a diagram of toaster face A, person, and toaster face B (similar to a worker-machine diagram). *Note:* Assume that the toaster already has the right temperature to toast at the starting time.

Problem MS2

Consider the following times and costs of a certain process:

Worker
Insert part	0.6 min
Remove part	0.4 min
Inspect part	0.6 min
Register part	0.2 min
Walk to the next machine	0.2 min
Salary	2 $/h

Machine
Processing time	4 min
Repayment	3 $/h

How many machines will the worker be able to run?

Problem MS3

In the packing section of a soap company, the machines are old, and they have a lot of problems. The enterprise managers are thinking of changing the machines, but they do not know how many they should purchase. Only one worker is assigned to that section, and the worker is responsible for a palletizer.

The new machines have a packaging time of 4.5 minutes. The machine repayment is 4 $/h. On the other hand, the worker (salary 2 $/h) should carry out the following tasks that take him or her the times shown in the following table (W = only the worker; M + W = machine and worker):

Task	Duration
Place the tip of plastic on the pallet (M + W)	1 min
Weld the end of the plastic (M + W)	0.2 min
Inspect the weld (W)	0.2 min
Label the pallet (W)	1 min
Walk to the following palletizer (W)	0.2 min

1. How many machines can the worker operate?
2. What is the minimum cost for each palletized unit?

Problem MS4

At a Disneyland resort, a new show based on the *Mulan* film is being produced. The performance consists of observing a juggler who rotates plates on some supports arranged in a circle. The objective of the show is to rotate, over a time still to be decided (but greater than 5 minutes), as many plates as possible.

The layout of the supports (arrange in a circle) is such that the juggler takes the same time in arriving at all the plate positions (4 seconds to go from one position to the next one). The plates can remain rotating without falling for 2.5 minutes. The time that it takes the juggler to place and rotate each plate is 3 seconds (the same time that takes to rotate the plate again so that it does not fall).

1. What is the maximum number of plates that the juggler will be able to rotate?
2. Why?

Problem MS5

The left worker-machine diagram in the following figure shows how a worker controls a machine. There is no free time for the worker, and the machine is stopped for 4 minutes in each cycle. This relationship can be changed by including an assistant, as shown in the worker-assistant-machine diagram on the right. The needed data are:

Product sale price:	2.4 $/unit
Salary of the worker:	3.6 $/h
Salary of the assistant:	1.8 $/h
Repayment of the machine:	10.2 $/h

WORKER	MACHINE
Prepare material (2 min.)	Idle time
Insert part (1 min)	OCCUPIED
Pack previous part (4 min.)	Processing part (4 min.)
Remove part (1 min)	OCCUPIED
Inspect part (2 min.)	Idle time

WORKER	ASSISTANT	MACHINE
Insert part (1 min)	Prepare material (2 min.)	OCCUPIED
Inspect previous part (4 min.)	Idle time	Processing part (4 min.)
Idle time		
Pack previous part (1 min)	Pack previous part (3 min.)	
Remove part (1 min)		OCCUPIED

Which of the two methods is more profitable?

Problem MS6

One worker runs four identical machines arranged in a circle, so the displacement time to each machine can be considered zero. The salary of the worker is 9 $/hour. Each machines has a repayment cost of 15 $/hour.

Each machine needs 15 minutes to process one piece, and it has an automatic feeder and extracting device. The feeder cannot accept the trailing piece from the conveyor belt until the worker removes the previous piece.

The piece adjustment must be exact, and as a consequence, the machine needs 1 minute to place the piece in the right position. Besides, owing to the high temperature of the finished pieces, the extracting device takes 4 minutes to remove the piece from the machine.

The worker uses 3 minutes to inspect each piece and an additional 3 minutes to register and pack the piece. The company works three shifts of 8 hours each.

1. How much money would it cost to invest in a refrigeration device that reduces the extracting time 3 minutes so that the cost of each piece decreases by 10 percent? How much would the daily production increase with this new device?
2. How much does the cost of each piece decrease if the company invests in a digital caliper for the worker that reduces the inspection time by half? The repayment cost of the caliper is 0.12 $/hour. How much does the daily production increase? *Note:* Assume a stable state.

Problem MS7

The worker-machine diagram in the following figure shows the manufacturing process for a wooden seat leg.

The company managers want to increase production. They do not know what is better: (1) the worker running two identical machines or (2) placing another worker in parallel with another identical machine. The company works a shift of 8 hours daily, and the costs associated with this decision process are:

Salary of a worker	3 $/h
Repayment of one machine	10.2 $.h

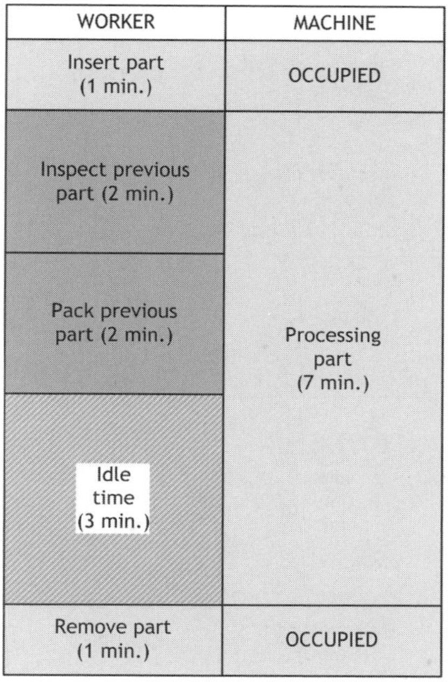

WORKER	MACHINE
Insert part (1 min.)	OCCUPIED
Inspect previous part (2 min.)	Processing part (7 min.)
Pack previous part (2 min.)	
Idle time (3 min.)	
Remove part (1 min.)	OCCUPIED

1. Which of the two methods is better from the cost point of view?
2. From a daily maximum production point of view?
3. Which option should the company choose?

Problem MS8

The worker-machine diagram in the following figure shows the manufacturing process for a wooden seat leg.

The company managers want to increase production. They do not know what is better: (1) placing another worker in parallel with another identical machine or (2) placing only one worker with two machines but including a *poka-yoke* for automatic product inspection in each machine (*poka-yoke* does not increase machine cycle). The company works a shift of 8 hours daily, and the costs associated with this decision process are

Salary of a worker	3 \$/h
Repayment of one machine	10.2 \$/h
Repayment of one *poka-yoke*	0.3 \$/h

WORKER	MACHINE
Insert part (1 min.)	OCCUPIED
Inspect previous part (2 min.)	
Pack previous part (2 min.)	Processing part (7 min.)
Idle time (3 min.)	
Remove part (1 min.)	OCCUPIED

1. Which one of the two methods is better from the cost point of view?
2. From a daily maximum production point of view?
3. Which option should the company choose? *Note:* Assume a stable state.

Problem MS9

Macario is an engineering student who wants to earn some money working in a fast-food restaurant. The following table shows the tasks involved in the process:

Macario
Open a potatoe bag	1 min
Empty bag into basket of deep fryer	2 min
Remove the basket from the deep fryer when fries are done	2 min
Empty the basket of fried potatoes	0.5 min
Walk to the following deep fryer	0.5 min
Prepare a portion of fries	0.5 min

Deep fryer
Fry a bag of potatoes	12 min

The salary is proportional to the work that Macario carries out, and he earns 60 cents for each bag of potatoes that he fries and 3 cents for each portion of fries he prepares (in the case of not preparing the portions, a partner will prepare them). Each deep fryer has only one basket, which Macario has to empty into a big container before being able to begin the cycle again.

How much money can Macario earn monthly if he works 8 hours a day, 20 days a month? *Note:* For each bag, 15 portions can be obtained, and there are four deep fryers (not all of them have to be used at all times, and the company does not lose money if some of them are stopped). If Macario does not take the basket out from the deep fryer, the potatoes will burn.

Problem MS10

The working cycle for a process has been studied using a stopwatch. The following table shows the tasks and the average times obtained from the chronometer:

Worker	
Place piece	2 min, 15 s
Remove piece	3 min, 25 s
Measure and pack	1 min, 5 s
Machine	
Process	10 min

Assume 10 percent of extra time for personal allowances and an activity factor of 1.2. The worker commented that after manufacturing 1000 pieces, the tool should be changed, which takes 25 minutes (standard time).

1. How many machines can the worker run if his or her salary is 6 $/hour and the machine repayment is 9 $/hour?
2. What is the cost of each finished piece? *Note:* The displacement time from one machine to another can be ignored.

Problem MS11

The worker-machine diagram in the following figure shows the manufacturing process for a plastic component.

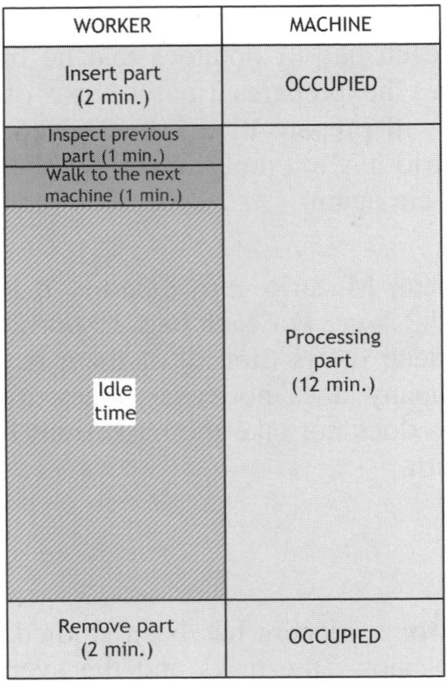

WORKER	MACHINE
Insert part (2 min.)	OCCUPIED
Inspect previous part (1 min.) Walk to the next machine (1 min.) Idle time	Processing part (12 min.)
Remove part (2 min.)	OCCUPIED

The company managers want to improve the process. They do not know which is better: (1) placing two workers each running two different machines or (2) placing one worker with a specified number of machines. The company works one shift of 8 hours daily, and the costs associated with this decision process are

Salary of a worker	5 $/h
Repayment of one machine	9 $/h

1. Which of the two methods is better from a cost point of view?
2. From a daily maximum production point of view?
3. Which option should the company choose?

Problem MS12

The worker-machine diagrams in the following figure show the manufacturing processes for two components.

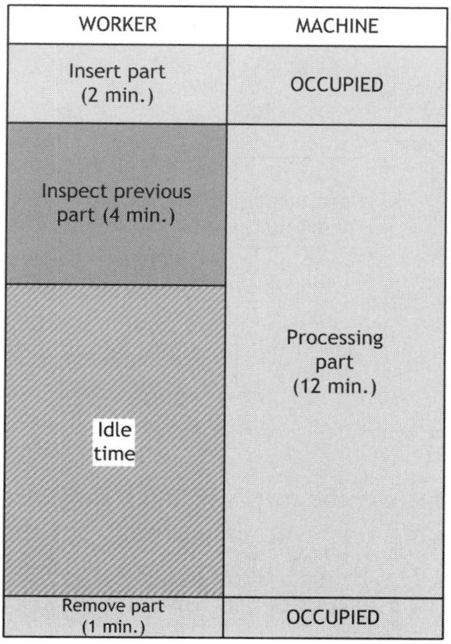

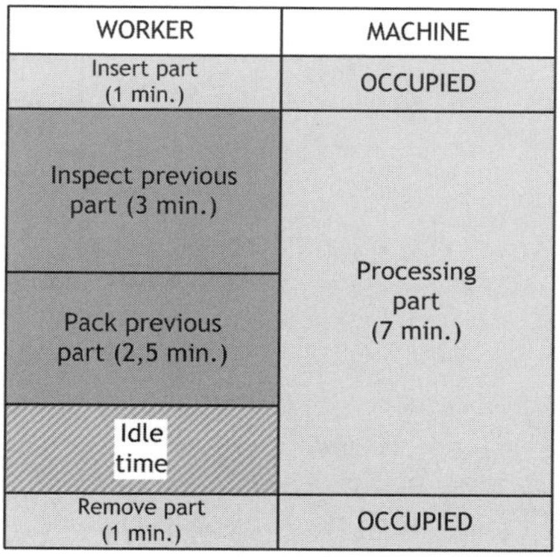

WORKER	MACHINE
Insert part (2 min.)	OCCUPIED
Inspect previous part (4 min.)	Processing part (12 min.)
Idle time	
Remove part (1 min.)	OCCUPIED

WORKER	MACHINE
Insert part (1 min.)	OCCUPIED
Inspect previous part (3 min.)	Processing part (7 min.)
Pack previous part (2,5 min.)	
Idle time	
Remove part (1 min.)	OCCUPIED

The company managers want to improve this process. They do not know which is better: (1) placing one worker on each machine or (2) placing one worker running both machines (in this case, the displacement time to each machine of 15 seconds should be considered). The company works one shift of 8 hours daily, and the costs associated with this decision process are

Salary of a worker	6 $/h
Repayment of one machine	10.2 $/h

1. Which of the two methods is better from a cost point of view?
2. From a daily maximum production point of view?
3. Which option should the company choose?

Problem MS13

The following figure represents the assembly process for the parts that cell B24 manufactures. The machine carries out automatic assembly of the components that come from M1 and M2, and its repayment cost can be ignored.

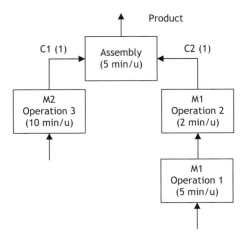

The company managers want to improve the process, and they are considering two alternatives: (1) placing one worker carrying out the tasks of M2 and another worker at M1 or (2) placing two workers carrying out the whole process, each one in a cell. (In the second case, two machines of each type would be needed)

At the M2 process, it takes 1 minute to introduce the part, 1 minute to extract it, and 2 minutes to inspect it. At M1, the change from operation 1 to operation 2 takes 1 minute.

The company works one shift of 8 hours daily, and the costs associated with this decision process are

Salary of a worker	4.81 $/h
Repayment of one machine	9.02 $/h

1. Which of the two methods is better from a cost point of view?
2. From a daily maximum production point of view?
3. Which option should the company choose?

Problem MS14

The following figure represents the assembly process for the parts that cell B24 manufactures. The machine carries out automatic assembly of the components that come from M1 and M2, and its repayment cost can be ignored.

The company managers want to improve the process, and they are analyzing two alternatives: (1) placing one worker carrying out the tasks of M2 and another worker at M1 or (2) placing two workers carrying out the whole process, but each one in a cell. (In this second case, two machines of each type would be needed.)

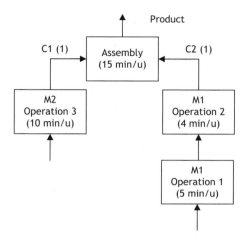

At the M2 process, it takes 1 minute to introduce the part, 1 minute to extract it, and 2 minutes to inspect it. At M1, the change from operation 1 to operation 2 takes 1 minute.

The company works one shift of 8 hours daily, and the costs associated with this decision process are

Salary of a worker	4.81 $/h
Repayment of one machine	9.02 $/h

1. Which of the two methods is better from a cost point of view?
2. From a daily maximum production point of view?
3. Which option should the company choose?

Problem MS15

A family has two electric grills to roast T-bones. Because the T-bones to be roasted are big, only one T-bone fits on each grill at a time, and it is necessary to use some grippers to turn the T-bone with both hands.

The family wants to roast three T-bones in the shortest time possible. The times needed to carry out the process are as follows:

Put on a T-bone:	10 s
Roast a T-bone side:	25 s
Turn/flip the T-bone:	15 s
Remove the T-bone:	10 s

1. How much time is required to toast the three T-bones on both sides?
2. Show a diagram entitled, "grill 1–grill 2" to prove it (similar to a worker-machine diagram). *Note:* Assume that the grills already have the right starting temperature to roast at the starting time. If the T-bone is not flipped to the other face or taken off when it is suppose to be (overcooked), the family will not accept it.

Problem MS16

The following figure represents the assembly process for the parts that cell B24 manufactures. The machine carries out automatic assembly of the components that come from M1 and M2, and its repayment cost can be ignored.

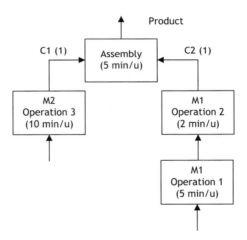

The company managers want to improve the process, and they are analyzing two alternatives: (1) placing one worker carrying out the tasks of M2 and another worker at M1 or (2) placing two workers carrying out the whole process, each one in a cell. (In this second case, two machines of each type would be needed).

In the M2 process, it takes 1 minute to introduce the part, 1 minute to extract it, and 2 minutes more to inspect it. In M1, the change from operation 1 to operation 2 takes 1 minute.

The company works one shift of 8 hours daily, and the costs associated with this decision process are

Salary of a worker	4.81 $/h
Repayment of one machine	9.02 $/h

Show the worker-machine diagram for both situations, assuming a stable state.

Problem MS17

Maria is an engineering student who wants to earn some money working in a well-known fast-food restaurant. The salary is proportional to the work that Maria carries out, and she earns 60 cents for each bag of potatoes she fries and 3 cents for each portion of potatoes she prepares (in the case of not preparing the portions, a partner will prepare them).

Each deep fryer only has one basket, and Maria has to empty it into a big container before being able to start the cycle again. The tasks involved in the process are as follows:

Maria
Open a potatoe bag	1 min
Empty the bag into the basket of the deep fryer	2 min
Remove the basket from the deep fryer when done	2 min
Empty the basket of fried potatoes	0.5 min
Walk to the next deep fryer	0.5 min
Prepare a portion of fries	0.5 min

Deep fryer
Fry a bag of potatoes	12 min

Show using a diagram entitled, "Maria–deep fryer" (similar to a worker-machine diagram) the situation in which Maria earns the most money. *Note:* From each big bag, 15 portions can be obtained, and there are four deep fryers (not all of them have to be used at all times, and the company does not lose money if some of them are stopped). If Maria does not remove the basket from the deep fryer on time, the potatoes will burn.

Problem MS18

Tom is an engineering student who wants to earn some money working in a well-known fast-food restaurant. The salary is proportional to the work that Tom carries out, and he earns 60 cents for each bag of potatoes he fries and 3 cents for each portion of potatoes he prepares (in the case of not preparing the portions, a partner will prepare them).

Each deep fryer has *two* baskets, but the deep fryer is only able to fry one basket in each working cycle.

The tasks involved in the process are as follows:

Tom
Open a bag of potatoes	1 min
Empty bag into a basket	1 min
Place the basket in the deep fryer	1 min
Remove the basket from the deep fryer when done	2 min
Empty the basket of fried potatoes	0.5 min
Walk to the next deep fryer	0.5 min
Prepare a portion fries	0.5 min

Deep fryer
Fry a bag of potatoes	12 min

1. How much money can Tom earn monthly if he works 8 hours a day, 20 days a month?
2. Show in a diagram entitled "Tom–deep fryer" (similar to a worker-machine diagram) this situation. *Note:* For each big bag, 15 portions can be obtained, and there are four deep fryers (not all of them have to be used at all times, and the company does not lose money if some of them are stopped). If Tom does not remove the basket from the deep fryer on time, the potatoes will burn.

Problem MS19

The worker-machine diagram in the following figure shows the manufacturing process for a plastic component.

The company managers want to increase component production. They do not know which is better: (1) placing only one worker with two identical machines or (2) placing one worker with two identical machines but including an assistant to pack the component. The company works one shift of 8 hours daily, and the costs associated with this decision process are

Salary of a worker	5 $/h
Salary of the assistant	3 $/h
Repayment of one machine	10 $/h

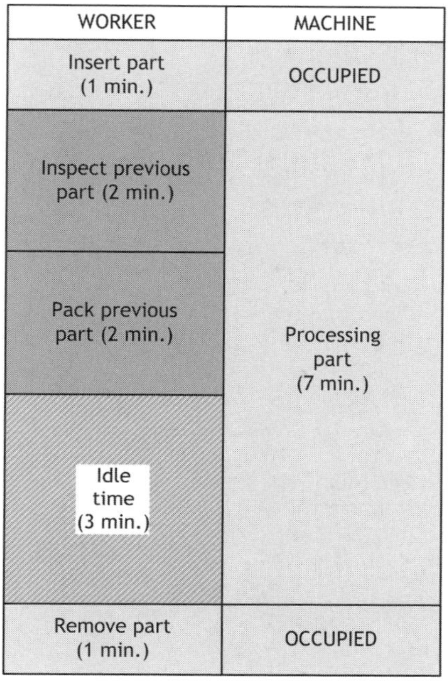

WORKER	MACHINE
Insert part (1 min.)	OCCUPIED
Inspect previous part (2 min.)	
Pack previous part (2 min.)	Processing part (7 min.)
Idle time (3 min.)	
Remove part (1 min.)	OCCUPIED

1. Which of the two methods is better from a cost point of view?
2. From a daily maximum production point of view?
3. Show the worker-machine diagram for both situations, assuming a stable state.

MACHINE-MACHINE DIAGRAMS

Problem MM1

The cell represented in the following figure is formed by three machines. The setup time between operations can be considered null, and both products have the same profit. The company works 5 days a week, 8 hours per shift, with two shifts per day.

1. Determine the maximum weekly production.
2. Calculate the cycle time.
3. Show the machine-machine diagram, assuming a stable state.

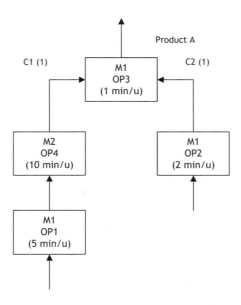

Problem MM2

The cell represented in the following figure is formed by two machines. The setup time between operations can be considered null, and both products have the same profit. The company works 7 days a week, 8 hours a day.

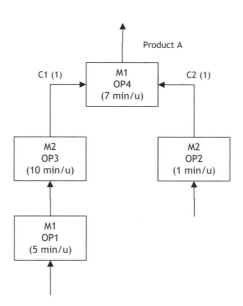

1. Determine the maximum weekly production.
2. Calculate the cycle time.
3. Show the machine-machine diagram, assuming a stable state.

Problem MM3

The cell represented in the following figure is formed by three machines. The setup time between operations can be considered null, and both products have the same profit. The company works 5 days a week, 8 hours per shift, with two shifts per day.

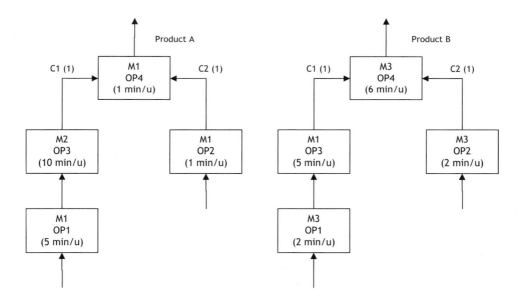

1. Determine the maximum weekly production.
2. Calculate the cycle time of both products.
3. Show the M2-M1-M3 diagram, assuming a stable state.

Problem MM4

The cell represented in the following figure is formed by two machines. The setup time between operations can be considered null, and both products have the same profit. The company works 5 days a week, 8 hours a day.

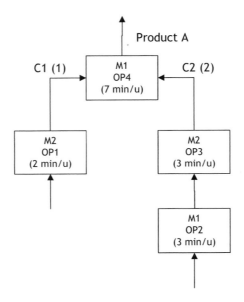

1. Determine the maximum weekly production.
2. Calculate the cycle time and the lead time
3. Show the machine-machine diagram, assuming a stable state.

Problem MM5

The cell represented in the following figure is formed by two machines. The setup time between operations can be considered null, and both products have the same profit. The company works 5 days a week, 8 hours a day.

The company uses the one-piece flow concept to manufacture the product. As a result, the batch size is one part. Besides, the part must be in the machine to start the setup process. The setup time in M2 is 1 minute, and the setup time in M1 (which depends on the operation sequence) is shown in the following table:

M1	Operation 1	Operation 3	Operation 4	Operation 6
Operation 1	—	2	3	2
Operation 3	2	—	1	3
Operation 4	3	1	—	1
Operation 6	2	3	1	—

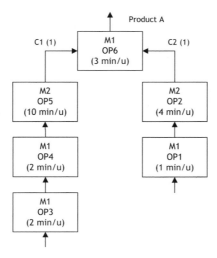

1. Determine the maximum weekly production.
2. Calculate the cycle time.
3. Show the machine-machine diagram, assuming a stable state.

Problem MM6

The cell represented in the following figure is formed by three machines. The setup time between operations can be considered null, and both products have the same profit. The company works 5 days a week, 8 hours a shift, with two shift per day.

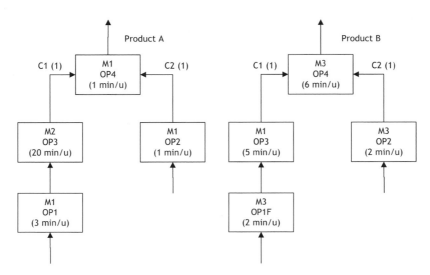

1. Determine the maximum weekly production.
2. Calculate the cycle time of both products.
3. Show the M2-M1-M3 diagram, assuming a stable state.

Problem MM7

The cell represented in the following figure is formed by three machines. The setup time between operations can be considered null, and both products have the same profit. The company works 5 days a week, 8 hours a shift, with two shifts per day.

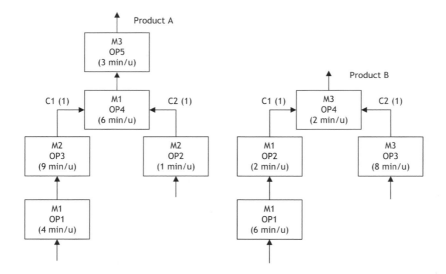

1. Determine the maximum weekly production.
2. Calculate the cycle time of both products.
3. Show the M2-M1-M3 diagram, assuming a stable state. Reorder the diagram to show the minimum lead time if necessary.

Problem MM8

The cell represented in the following figure is formed by three machines. The setup time between operations can be considered null, and both products have the same profit. The company works 5 days a week, 8 hours a shift, with two shifts per day.

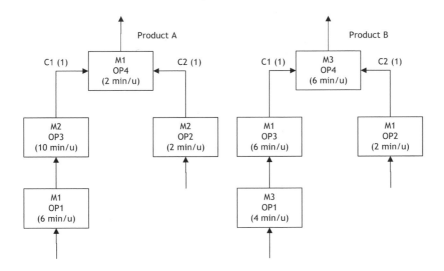

1. Determine the maximum weekly production.
2. Calculate the cycle time of both products.
3. Show the M2-M1-M3 diagram, assuming a stable state. Reorder the diagram to show the minimum lead time if necessary.

Problem MM9

The cell represented in the following figure is formed by three machines. The setup time between operations can be considered null, and both products have the same profit. The company works 5 days a week, 8 hours a shift, with two shifts per day.

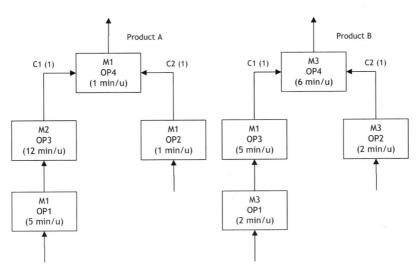

1. Determine the maximum weekly production.
2. Calculate the cycle time of both products.
3. Show the M2-M1-M3 diagram, assuming a stable state.

Problem MM10

The cell represented in the following figure is formed by three machines. The setup time between operations can be considered null, and both products have the same profit. The company works 5 days a week, 8 hours a shift, with two shifts per day.

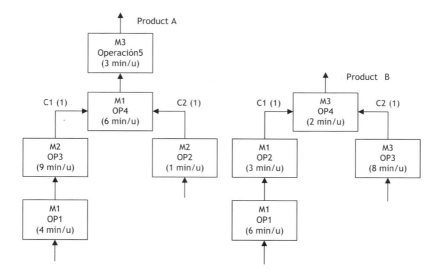

1. Determine the maximum weekly production.
2. Calculate the cycle time of both products.
3. Show the M2-M1-M3 diagram, assuming a stable state. Reorder the diagram to show the minimum lead time if necessary.

Problem MM11

The cell represented in the following figure is formed by two machines. The setup time between operations can be considered null, and both products have the same profit. The company works 5 days a week, 8 hours a shift, with two shifts per day.

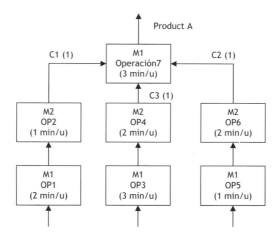

1. Determine the maximum weekly production.
2. Calculate the cycle time.
3. Show the machine-machine diagram, assuming a stable state.

Problem MM12

The cell represented in the following figure is formed by three machines. The setup time between operations can be considered null, and both products have the same profit. The company works 5 days a week, 8 hours a shift, with two shifts per day.

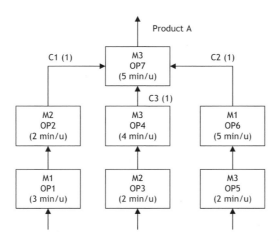

1. Determine the maximum weekly production.
2. Calculate the cycle time.
3. Show the M2-M1-M3 diagram, assuming a stable state.

Problem MM13

The cell represented in the following figure is formed by three machines. The setup time between operations can be considered null, and both products have the same profit. The company works 5 days a week, 8 hours a shift, with two shifts per day.

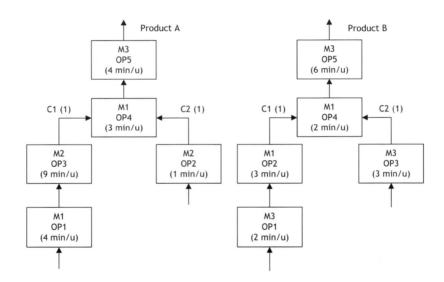

1. Determine the maximum weekly production.
2. Calculate the cycle time of both products.
3. Show the M2-M1-M3 diagram, assuming a stable state.

Problem MM14

The cell represented in the following figure is formed by three machines. The setup time between operations can be considered null, and both products have the same profit. The company works 5 days a week, 8 hours a shift, with two shift per day.

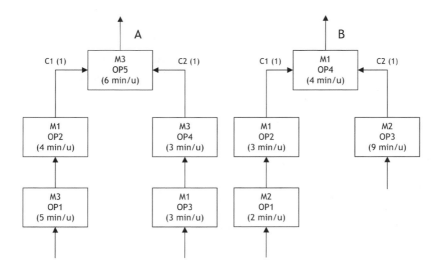

1. Determine the maximum weekly production.
2. Calculate the cycle time of both products.
3. Show the M3-M1-M2 diagram, assuming a stable state. Reorder the diagram to show the minimum lead time if necessary.

Problem MM15

The cell represented in the following figure is formed by three machines. The setup time between operations can be considered null, and both products have the same profit. The company works 5 days a week, 8 hours a shift, with two shifts per day.

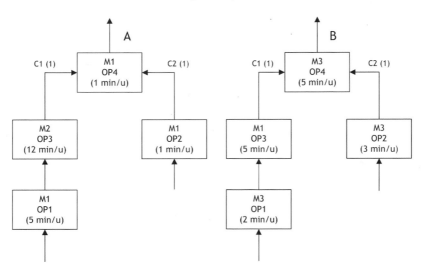

1. Determine the maximum weekly production.
2. Calculate the cycle time of both products.
3. Show the M2-M1-M3 diagram, assuming a stable state. Reorder the diagram to show the minimum lead time if necessary.

NUMERIC PROBLEM SOLUTIONS

Continuous Improvement Tools

Problem IT1

1. 58.12 percent
2. Quality = 0.778; performance = 0.854; availability = 0.875
3. Quality

Problem IT2

1. 41.8 percent
2. Quality = 0.715; performance = 0.665; availability = 0.875
3. Performance

Problem IT3

1. 44.5 percent
2. Quality = 0.715; performance = 0.719; availability = 0.897
3. Quality

Problem IT4

1. Initial OEE = 0.767; final OEE = 0.771
2. Yes, because, in this case, SMED achieves flexibility.
3. Because SMED improves quality, eliminating trials.

Problem IT5

1. Initial OEE = 0.849; final OEE = 0.72
2. Productivity has increased, but performance rate has decreased.

Problem IT6

1. Initial OEE = 0.809; final OEE = 0.864
2. OEE has increased, but production has decreased. Reduce the duration of daily planned stops.

Problem IT7

1. Initial OEE = 0.731; final OEE = 0.713
2. OEE has decreased, but with *poka-yoke* and source inspection implementation in the beginning, the defect rate will increase. They were worthwhile.

Problem IT8

1. 61.42 percent
2. Quality = 0.787; performance = 0.892; availability = 0.875
3. Quality

Facilities Layout

Problem FL1

1. Graphic answer
2. Sections 1 and 2 will be interchanged.

Problem FL2

1. Look at the theory.
2. Graphic answer

Problem FL3

1. Graphic answer

Problem FL4

1. Graphic answer

Cellular Layout

Problem CL1

1. ST = 42.53 min
2. Standard production 1.41 parts/h

Problem CL2

1. ACF/DGI/B/EHJ/KL
2. B/AD/CEG/IFH/JKL
3. ACF/B/DH/GEIJ/KL

Problem CL3
1. AD/BE/IC/FHG/JK/L
2. H = 17 s; F = 180 s; WIP = 3 units
3. Graphic answer

Problem CL4
1. A/CB/D/GF/—/I/EJ/H/KL
2. H = 13 min; F = 60 min; WIP = 5 units
3. Graphic answer

Problem CL5
1. A/C/BF/E/DG(double)/HIJ
2. H = 6 min; WIP = 7 units
3. Graphic answer

Problem CL6
1. Graphic answer
2. EDC/IHF/AGBJ; yes.
3. Mmin = 3
4. HR = 1.85%

Problem CL7
1. A/B/CDEG/HFI
2. H = 11 min; F = 32 min; WIP = 4 units
3. Graphic answer

Problem CL8
1. A/BE/CF/JD/GIH/KL/M
2. H = 9 s; F = 210 s; WIP = 3.5 units
3. Graphic answer

Problem CL9
1. A/CDB/EFGH(double)/I/J
2. H = 5 min; F = 40 min; WIP = 4 units
3. Graphic answer
4. It is not recommendable.

Problem CL10

1. T3/T1T2/T4T5T6/V1V2/V3V4/V5V6/V7MF1
2. H = 34 min
3. Graphic answer

Problem CL11

1. A/CD/BEG/F/HIJ/KLM
2. Two workers
3. H = 5 min; F = 120 min; WIP = 3 units
4. Graphic answer

Problem CL12

1. A/BE/CDGFH(double)/JI/KL/M
2. H = 12 s; F = 270 s; WIP = 4.5 units
3. Graphic answer

Problem CL13

1. AC/B/DE/FGH(double)/IK/JL/M
2. H = 6 s; F = 150 s; WIP = 5 units
3. Graphic answer

Problem CL14

1. A/D/CB/FEG/H/I
2. H = 23 min; F = 70 min; WIP = 3.5 units
3. Graphic answer

Problem CL15

1. Two parallel lines with a demand of 40 units
2. ACF/B/DH/GEIJ/KL
3. H = 5

Problem CL16

1. Two lines → B/AC/DE/GIFJ/H/K/L
2. WIP = 3.5 units on each line

Problem CL17

1. A/C/BF/E/DG/—/HIJ
2. WIP = 4 units; H = 16

Problem CL18
1. A/C/BF/E/DG/—/HIJ
2. WIP = 4 units; H = 16

Problem CL19
1. A/C/BF/ED/G/—/HIJ
2. WIP = 4 units; H = 4
3. Graphic answer

Problem CL20
1. A/DB/CF/E/HI/G/JK/L
2. WIP = 4 units; H = 5
3. Graphic answer

Problem CL21
1. A/BF/D/CE/GH/I/—/JK
2. WIP = 5 units; H = 4
3. Graphic answer

Problem CL22
1. ADB/E/CFG/IH/KJL (double)
2. H = 10
3. Graphic answer

Problem CL23
1. AC/BFE/D/G/H/IJ
2. WIP = 3.5 units; H = 0
3. Graphic answer

Maintenance

Problem MN1
1. Depending on the cost
2. Look at the theory.

Problem MN2
1. 87.5 percent
2. Look at the theory.

Problem MN3

1. Component B
2. Look at the theory.

Motion Study

Problem MS1

1. 135 s
2. Graphic answer

Problem MS2

Two machines

Problem MS3

1. Two machines
2. 0.48 $/unit

Problem MS4

1. 21 plates
2. Because if manage 22, the plates fall.

Problem MS5

1. Better with an assistant (8.4 $ profit/hour)

Problem MS6

1. Nothing
2. It is reduced 1.14 $/part

Problem MS7

1. Two workers in parallel
2. Two workers in parallel
3. Two workers in parallel

Problem MS8

1. Better with a *poka-yoke*
2. The same in both cases
3. Better with a *poka-yoke*

Problem MS9
855.3 $/month

Problem MS10
1. One worker and two machines
2. 3.56 $/part

Problem MS11
1. Both are the same
2. Better two workers with two machines each worker
3. Two workers with two machines each worker

Problem MS12
1. M1 product is more expensive, and M2 product is cheaper.
2. One worker on each machine
3. One worker on each machine

Problem MS13
1. Two workers in parallel
2. Two workers in parallel
3. Two workers in parallel

Problem MS14
1. Two workers in parallel
2. Two workers in parallel
3. Two workers in parallel

Problem MS15
1. 145 s
2. Graphic answer

Problem MS16
Graphic answer

Problem MS17
Graphic answer

Problem MS18

1. 883.2 $/month
2. Graphic answer

Problem MS19

1. One worker, two machines, and one assistant
2. One worker, two machines, and one assistant
3. Graphic answer

Machine-Machine Diagram

Problem MM1

1. 480 products
2. 12 min
3. Graphic answer

Problem MM2

1. 280 products
2. 12 min
3. Graphic answer

Problem MM3

1. 400 A with 400 B
2. 12 minutes for each product
3. Graphic answer

Problem MM4

1. 185 products
2. 13 min; 26 min
3. Graphic answer

Problem MM5

1. 150 products
2. 16 min
3. Graphic answer

Problem MM6

1. 240 A and 480 B
2. $tc_A = 20$ min; $tc_B = 10$ min
3. Graphic answer

Problem MM7

1. 267 A and 267 B
2. 18 min
3. Graphic answer

Problem MM8

1. 300 A and 300 B
2. 16 min
3. Graphic answer

Problem MM9

1. 400 A and 400 B
2. 12 min
3. Graphic answer

Problem MM10

1. 252 A and 252 B
2. 19 min
3. Graphic answer

Problem MM11

1. 533 products
2. 9 min
3. Graphic answer

Problem MM12

1. 436 A
2. 11 min
3. Graphic answer

Problem MM13

1. 320 A and 320 B
2. 15 min

3. Graphic answer

Problem MM14

1. 343 A and 343 B
2. 14 min
3. Graphic answer

Problem MM15

1. 400 A and 400 B
2. 12 min
3. Graphic answer

Index